Medical Emergencies Caused by Aquatic Animals

Vidal Haddad Junior

Medical Emergencies Caused by Aquatic Animals

A Biological and Clinical Guide to Trauma and Envenomation Cases

Second Edition

 Springer

Vidal Haddad Junior
Botucatu Medical School
Sao Paulo State University
Botucatu
São Paulo
Brazil

ISBN 978-3-030-72252-4   ISBN 978-3-030-72250-0   (eBook)
https://doi.org/10.1007/978-3-030-72250-0

This Springer imprint is published by the registered company Springer Nature Switzerland AG
The registered company address is: Gewerbestrasse 11, 6330 Cham, Switzerland

Dedicated to Adriana, Lorenzo, and Ariadne, gifts that life has given me.

Foreword

Hubiera sido, pues, excelente ocasión para practicar la obra de caridad más propia de nuestro tiempo: no publicar libros superfluos.

Ortega y Gasset

Over the past 22 years, the gods have given me the privilege of enjoying the friendship and companionship of two unique human beings, Rubens Brando, the "Rubão," and Vidal Haddad Jr. Both, each in his own way, with the nobility of character and sovereignty purposes that the ancient Greeks defined well with the word *Arete*. Rubão and Vidal: scientists, aristocrats of the spirit.

Rubens, with his Indians and his inordinate passion for snakes, left us long ago.

Vidal, with his tenacity and dedication, now rewards us with this text. I participated in his project from the early stages, when about 20 years ago, he headed up to the Santa Casa de Ubatuba to ask the staff of that hospital for the necessary collaboration to develop clinical observations of patients injured by marine animals.

The doctor Hector P. Froes, in the 1930s, in Bahia State, was the last Brazilian author to be internationally recognized for his observations on the venomous fish of Brazil.

The track was redone.

The book presented here, with magnificent iconographic material, portrays an experience that was lived with passion, in the day-to-day field work, on trips, in contact with patients, and in visits to places of where incidents occurred. It is up to the reader judge it.

This will certainly not be a superfluous book.

São Paulo, Brazil   João Luiz Cardoso

Acknowledgments

João Luiz Costa Cardoso

 the greatest contributor and supporter of this work
 "But who really knows, is not talking around …
 Remains quiet, waiting to be asked and generally
 solves the problem, since the silence is part of
 his wisdom."
 To my parents Vidal and Maria Stella, my siblings William and Cristiane,
my sister and brother-in-law Margareth and Paulo Cezar, and my nephews
 Manoel, Luzinete, and Julio Linuesa Peres
 Drs. Ricardo Cortes and Afrânio Borsatto
 (Santa Casa de Ubatuba, São Paulo State)
 Prof. Dr. Edmundo Ferraz Nonato, *in memoriam*
 (Oceanographic Institute of São Paulo)
 Dra. Neuza Lima Dillon, *in memoriam*, and colleagues of the Department
of Dermatology, FMB-UNESP
 Prof. Dr. Fabio Lang da Silveira, Prof. André Carrara Morandini, and Prof.
Álvaro Esteves Migotto
 (Zoology, IB-USP and Centro de Biologia Marinha, USP, São Sebastião
town)
 Prof. Dr. Juan Pedro Lonza Joustra
 (Iquique, Chile)
 Prof. Itamar Alves Martins
 (UNITAU, Taubaté, São Paulo)
 Prof. Ivan Sazima (UNICAMP, Museum of Zoology), Prof. Otto Bismarck
Fazzano Gadig (Universidade Estadual Paulista, SP), Prof. José Sabino
(Universidade Anhanguera, Mato Grosso do Sul, and Prof. Jansen Zuanon
(INPA – COBIO – Amazonas)
 Prof. Nélson Oliveira Henrique (Orthopedics and Traumatology,
University Hospital Getulio Vargas, Manaus, Amazonas) and Dr. Anoar
Samad (Urology Services of the Portuguese Beneficient Society of Manaus
and Adventist Hospital Manaus)
 Dra. Monica Lopes-Ferreira
 (Immunopathology Laboratory, Butantan Institute of São Paulo – for the
collaboration and conjunct works)
 Prof. Alejandro Solorzano
 (National Serpentarium, San Jose, Costa Rica)
 Prof. Jorge Luiz da Silva Nunes

(Universidade Federal do Maranhão)

Prof.ª Fernanda Maria Duarte do Amaral

(Universidade Federal Rural de Pernambuco)

Manoel Francisco de Campos Neto, MD

POLITEC, Cáceres Regional, Mato Grosso State, Brazil

Prof. Domingos Garrone Neto,

(Universidade Estadual Paulista, who shares with me an interest in potentially dangerous aquatic animals)

Prof. Joseph W. Burnett (University of Maryland, Baltimore), Prof. John Williamson, and Prof. Findlay Russell (University of Texas), *in memoriam* – for encouragement and support during the phases of this book

To the freshwater and marine fishermen of Brazil, whose information and wisdom constitute almost all the contents of this book

Contents

Introduction

Actually, there is a considerable interest in aquatic environments due to recreational and professional activities that take place in such environments. This fact can result in dangerous encounters between aquatic animals and humans. Humans while engaging in leisure activities fall prey to the lesions, traumas, and envenoming, which are caused by aquatic fauna as a means of natural defense [1–5]. The main victims of these injuries are swimmers, professional and sportive fishermen, surfers, and scuba divers [1–5].

Poisonous, venomous, and traumatogenic animals mainly cause human injuries. Poisonous animals are those that contain toxins that cause deleterious effects when they are ingested or when they come into contact with the victim. This type of envenoming is caused by animals such as the pufferfish, toads, and some beetles. Venomous are those animals that can inject toxins through an apparatus such as spines or stingers. These toxins have diverse effects and cause proteolysis, myotoxicity, hemotoxicity (mainly hemolysis), cytotoxicity, and neurotoxicity. The main aquatic animals that can cause emergency situations in humans are included in the Phyla Porifera (sponges), Cnidaria (anemones, corals, jellyfish, and Portuguese man-of-war), Annelida (marine worms), Mollusca (octopuses and *Conus* shells), Echinodermata (sea urchins, starfish, and sea cucumbers), Crustacea (crabs and mantis shrimp), and Chordata (fish and reptiles) [1–5].

Injuries caused by marine animals are predominant in the summertime, when the population of the coastal towns increases exponentially. Bathers constitute more than 90% of the victims and the incidence of this type of accident is 0.1%, or 1 in 1000 patients admitted to emergency rooms [1–5]. Of the victims, nearly 50% are bathers who are injured by sea urchins and present with traumatic or venomous injuries, 25% are bathers who are injured by cnidarians (jellyfish and Portuguese man-of-war), and 25% are fishermen who are injured by venomous fish such as catfish and stingrays [1–5].

It is important to know that in the early stages of the injuries, there will always be emergency situations for victims due to the pain and bleeding associated with the wounds. The pain can be very severe and systemic manifestations can be observed, such as compromise of the cardiac, respiratory, and urinary systems. *Occasionally, there is risk of death of the victim, especially if envenomation is caused by some cnidarians and fish.*

Brazil is one of the only countries in the world that offers statistics on the frequency and severity of *injuries* caused by aquatic animals. These data are useful for the prevention and provision of first aid care for these types of injuries, and such data were collected through three sources: through prospective work of the author, in colonies of fishermen and the coastline, and rivers and lakes

V. Haddad Junior, *Medical Emergencies Caused by Aquatic Animals*, https://doi.org/10.1007/978-3-030-72250-0_1

of Brazil. Data about injuries that occur bathers are also cataloged in these clinical series, and about 3000 injuries are observed in a period of nearly 20 years.

The profile of these injuries, envenomation, and poisonings is similar throughout the world, with small variations in certain areas: injuries caused by sea urchins are the most common, and these injuries are commonly registered in areas, such as beaches, where these echinoderms are frequently found. This is followed by injuries caused by cnidarians and fish (traumatic and venomous) and by poisoning due to consumption of fish and seafood, which occurs especially in, but not limited to, tropical areas.

Figure 1.1 shows the distribution of 144 accidents caused by aquatic animals observed by the author in 18 months in Ubatuba town, Southeast Coast of Brazil. The curve of distribution of patients clearly shows peaks in the summer periods, during which period the town's population increases about ten times. In most of the situations, during various seasons, the population of animals that cause *injuries* does not decrease; however, an increase in the number of cases recorded is due to the lack of

information on the encounters between the bathers and these animals. Of the victims, more than 90% were swimmers, and the incidence of this type of injury is 0.1% or 1 in 1000 cases admitted to the emergency room; this increase in number is due to the reason that a holiday season can cater up to nearly 5000 people in a single day in beaches.

Figure 1.2 shows the cases observed by the Fire Department (17 GB – Salvamar Paulista) in a large area of the North Coast of São Paulo State from 1997 to 2001. Almost all the calls to the department were regarding the injuries caused by cnidarians and venomous fish, and the injuries caused by sea urchins – which are responsible for about 50% according to the author's study – are not represented in statistical data on the injuries acquired by lifeguards. The probable reason for this is that the pain caused by the spikes of *sea urchins* is moderate and allows the mobility of the victim to the hospital. Envenomations by cnidarians and venomous fish are very painful and require assistance from lifeguards with regard to primary care and to remove or shift patients to nearby hospitals. In general, however, the data are coincident. The

Fig. 1.1 Distribution of 144 patients from January of 1997 to June of 1998

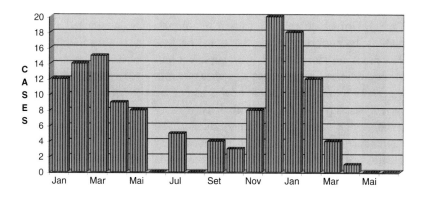

Fig. 1.2 Incidence of injuries in the São Paulo State, Brazil

	Cnidarian	Catfish	Fish (bites)	Others
GUARUJÁ town	14	3	2	1
SANTOS city	7	18	0	4
S.VICENTE city	0	2	1	1
BERTIOGA town	0	1	0	1
P.GRANDE town	1	5	1	0
MONGAGUÁ town	1	2	0	0
ITANHAÉM town	1	1	0	0
PERUÍBE town	4	1	0	0
S.SEBASTIÃO town	3	0	0	1
CARAGUÁ town	5	4	0	0
UBATUBA town	4	4	0	2
TOTAL	40	41	4	10

majority of the envenomation by fish *was* caused by catfish. This occurs when bathers accidentally step on these fish that were discharged in the beach sand by fishermen.

According to the Antivenin Information Center of Bahia (CIAVE) in 2001, localized in the Northeast Coast of Brazil, the injuries by cnidarians were responsible for the majority of calls in the area (54.89%), but there was a high incidence of injuries caused by niquins or toadfish (*Thalassophryne* sp.). This venomous fish is very common in estuarine areas of the North and Northeast regions. Most accidents occur in the warmer months of the year (January, February, and March), as indeed occurs on the coast of the Southeast, and this fact is related to the increased number of bathers on the beaches. In this series we did not record the traumatic accidents by sea urchins that cause traumatic injuries. Pain was the main symptom, with edema and erythema. Also poisoning by consumption of pufferfish meat, a common habit in Bahia State, is worth noting.

An extensive and recent work of the Ministry of Health of Brazil (2015) presents important epidemiological data: most of the injuries reported were caused by marine and freshwater stingrays and, in general, by venomous fish and cnidarians (Fig. 1.3). These data were obtained from the notifications by health teams. Based on these data, it has been assumed that all the causative animals were venomous, especially cnidarians and fish such as catfish and stingrays, but again the injuries by sea urchins were excluded (which are traumatic injuries and more commonly reported in prospective studies on bathers) for an understandable reason: most accidents in Brazil are caused by the black sea urchin (*Echinometra lucunter*), which does not provoke envenomations or severe pain, thereby not demanding immediate transport to emergency centers.

Aquatic animals	2007		2008		2009		2010		2011		2012		2013			
	N	%	N	%	N	%	N	%	N	%	N	%	N	%	N	%
Stingrays	192	42,7	331	58,4	387	67,5	404	76,4	429	77,7	624	82,5	473	68,4	40	68,9
Cnidarians	114	25,3	127	22,4	75	13,1	26	4,9	43	7,8	47	6,2	108	15,6	540	13,1
Toadfish	18	4,0	14	2,5	26	4,5	33	6,2	29	5,3	26	3,4	36	5,2	182	4,4
Catfish	9	2,0	6	1,1	6	1,0	9	1,7	4	0,7	14	1,9	23	3,3	71	1,7
Sea *urchins*	3	0,7	2	0,4	3	0,5	2	0,4	2	0,4	1	0,1	4	0,6	17	0,4
Other	114	25,3	87	15,3	76	13,3	55	10,4	45	8,2	44	5,8	48	6,9	469	11,4
Brazil	450	100	567	100.1	573	99.9	529	100	552	100.1	756	99.9	692	100		

Fig. 1.3   Injuries caused by aquatic animals: Brazil, 2007–2013. (Data from SINAN/SVS/Health Ministry)

This work intends to provide information on the main animals that cause these injuries (all these injuries fall under the same category across the world) and the actual treatment utilized for the trauma and envenomations described.

At the time of preparing this second edition, we are experiencing the pandemic of the COVID-19 virus, with catastrophic repercussions on the human species, all over the planet. As a result, the numbers of injuries by aquatic animals have decreased due to the need to avoid agglomerations and the accelerated transmission of the virus. We should not deceive ourselves, however. This temporary decline may lead to an increase in the occurrence of trauma and envenomation caused by aquatic animals, because once the crisis is over, with the advent of a vaccine, there will be a much increased demand for aquatic leisure areas, both freshwater and marine. Certain populations of marine animals (cnidarians) exponentially increase at certain times of the year, and this fact can be associated with the frequency of encounter that takes place between aquatic animals and humans; for example, we can expect an increase in the encounter between humans and jellyfish in summer after the pandemic. In addition, certain outbreaks of injuries such as those caused by piranhas and stingrays in freshwater leisure areas also became more common, as the restricted areas were again opened for recreational activities.

More than ever, we need to understand that human is not the owner of the planet; he is just another species that depends on others to survive. The extermination of wild animals brought us the coronavirus. Still, several viruses such as Ebola are found in cave-dwelling bats. Aggressions to the aquatic environments have resulted in the following: increased poisonings due to dinoflagellates, red tides, and a bloom in the population of cnidarians, to mention a few. We have set up a time bomb. Will there be time to disarm?

References

1. Haddad Jr V. Avaliação Epidemiológica, Clínica e Terapêutica de Acidentes Provocados por Animais Peçonhentos Marinhos na Região Sudeste do Brasil (thesis). São Paulo (SP): Escola Paulista de Medicina, 1999. 144 pp.
2. Haddad V Jr. Atlas de animais aquáticos perigosos do Brasil: guia médico de diagnóstico e tratamento de acidentes (Atlas of dangerous aquatic animals of Brazil: a medical guide of diagnosis and treatment). São Paulo: Editora Roca; 2000. 148 pp
3. Haddad V Jr. Animais aquáticos de importância médica. Rev Soc Bras Med Trop. 2003;36:591–7.
4. Haddad V Jr. Animais Aquáticos Potencialmente Perigosos do Brasil: Guia médico e biológico (Potentially Dangerous aquatic animals of Brazil: a medical and biological guide). São Paulo: Editora Roca; 2008. 268 pp
5. Haddad V Jr, Lupi O, Lonza JP, Tyring SK. Tropical dermatology: marine and aquatic dermatology. J Am Acad Dermatol. 2009;61:733–50.

Phylum Porifera (Marine and Freshwater Sponges)

Sponges are simple metazoans and are classified under Phylum Porifera of the Animal Kingdom. Sponges have a hollow, circular body, which filters the water of the environment retaining the food. Spongin is a protein that forms the main part of the body and it consists of a "skeleton" of hard spike-like structures that are made up of calcium carbonate and silica. Additionally, the sponges present an irritating and possibly venomous slime on the surface of the body [1–5]. Various species of marine and freshwater sponges are associated with the skin and ophthalmic lesions in humans, but the main marine species involved are *Neofibularia* sp., *Tedania ignis* (the fire sponge), and *Microciona prolifera*, the red sponge (Figs. 2.1 and 2.2).

A localized skin eruption with an eczematous pattern is observed in a victim who comes in contact with marine sponges (Figs. 2.3 and 2.4). The onset is fast (1–3 hours) and the itch/burning sensation is intense. The dermatitis heals after about 2 weeks. The most common location is the hand palms, and actually dermatitis is only observed in biologists and in those who collect sponges for commercial uses. The manifestations are exclusively local and there are no systemic complications [1–5]. The slime can be removed with soap and clean water in recent contacts. The spikes present in the injured area

can be removed by using adhesive tapes. Late inflammation at the point of the penetration of the spikes and contact with the slime can be controlled with the use of corticosteroid creams and cold compresses [2, 6].

Freshwater sponges cause human skin lesions via a mechanism similar to that observed in marine sponges [4]. The most important class that causes lesions is Demospongiae, but there are other classes associated with cutaneous and ocular lesions. Freshwater sponges are found in the lakes and tributaries of rivers in the Amazon region, and they also are found in lakes in Savannah-like areas ("cerrados") in the Central and Southeast Brazil and in freshwater pools around the world (Figs. 2.5 and 2.6). In some regions of the Brazilian Amazon and the "cerrados," indigenous people call the freshwater sponges "cauxi" or "pó-de-mico" (monkey powder) due to the pruritus that is caused when people come in contact with the spikes.

Unlike the localized lesions caused by marine sponges, dermatitis associated with freshwater sponges is disseminated. The lesions are itchy erythematous papules presenting central vesicles at the point of the penetration of the spike and secondarily we can observe exulcerations, crusts, and bacterial infections (Fig. 2.7) [4]. The reason for the difference in the distribution of lesions is that the spikes are dispersed in the water of ponds and rivers and penetrate the uncovered skin and mucous membranes of bathers and riverine work-

Fig. 2.1 Fragment of a red marine sponge on a beach. Touching the sponge can provoke severe dermatitis in bathers. (Photo: Vidal Haddad Junior)

ers. The inflammation caused by spikes can manifest itself in the victims' eyes, causing serious eye damage and cases of blindness due to chronic inflammation. This serious complication was associated with spike suspensions in swimmers in recreational areas and in fishermen who dive to remove their nets. This happens mainly in areas of water with little movement and subject to periods of drought. At this moment, the freshwater sponges die and when the rains return, the ponds fill up and the spikes are suspended in the water, causing dermatitis and ophthalmopathies [7].

The treatment for injuries caused by marine and freshwater sponges is effective. If the manifestations are disseminated and severe (showing intense pruritus and marked inflammation), it is a fundamental act in the inflammatory process to use systemic corticosteroids, that is, 30–40 mg of

prednisone per day for about a week, with gradual withdrawal. If the eruption is localized (marine sponges) or disseminated but mild or moderate, application of topical corticosteroids and use of oral antihistamines will control the dermatitis. This is a common problem in some regions, but additional studies – including proofs for the etiology of the process and for histopathological analysis (i.e., visualization of the spikes present in the injured areas) – are lacking. Regarding ocular lesions caused by freshwater sponges, there are studies about epidemic uveitis and leukomas in riverside people associating the lesions with the presence of spikes of two freshwater sponges (*Drulia uruguayensis* and *D. ctenosclera*). In these cases, the diagnosis was confirmed by histopathology, demonstrating the etiology of ophthalmic process [7].

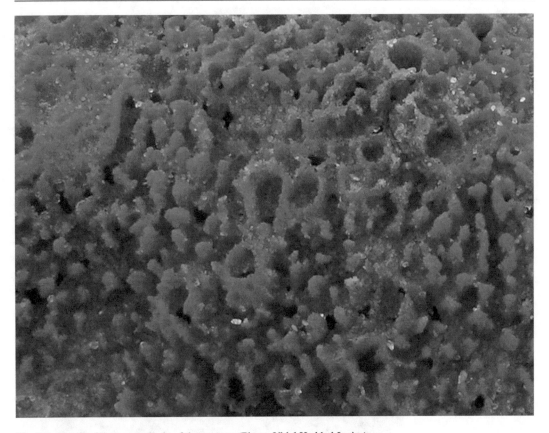

Fig. 2.2 Detail of the porous body of the sponge. (Photo: Vidal Haddad Junior)

Fig. 2.3 Eczematous plaques caused by contact with marine sponge in a biologist who collected sponges for research purposes. (Photo: Vidal Haddad Junior)

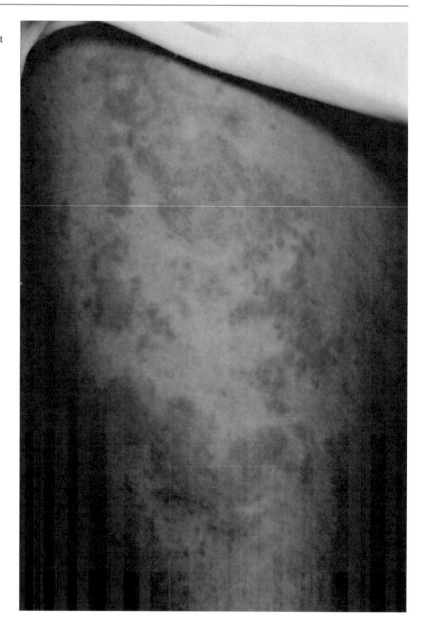

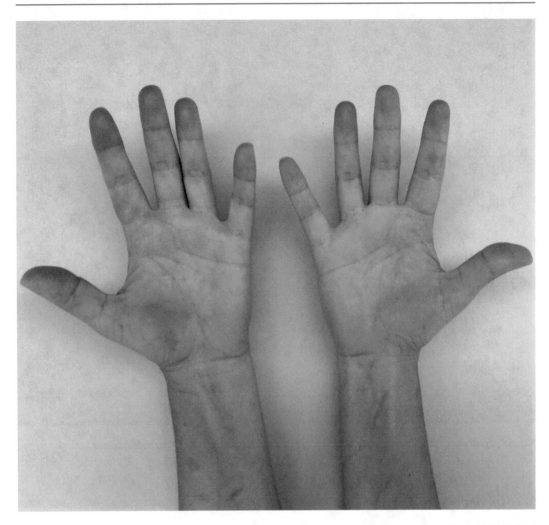

Fig. 2.4 Important inflammatory lesions observed in a patient that was collected marine sponges. (Photo: Vidal Haddad Junior)

Fig. 2.5 Freshwater sponge adhered to a branch of the sub-aquatic vegetation. (Photo: Vidal Haddad Junior)

Fig. 2.6 Detail of the body of the freshwater sponge, composed of organic filaments and siliceous and calcareous spicules. (Photo: Vidal Haddad Junior)

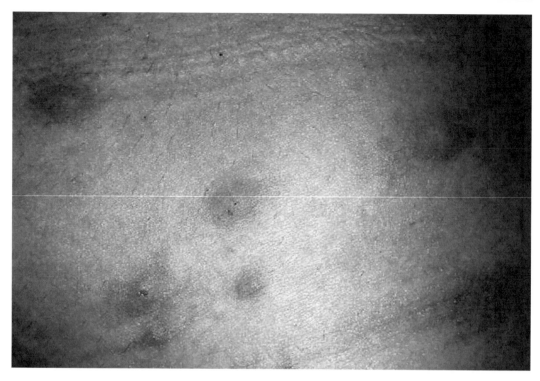

Fig. 2.7 Erythematous papules caused by the body spicules of a freshwater sponge. The eruption is disseminated once the spicules are in suspension in the water of rivers and lakes. (Photo: Vidal Haddad Junior)

Phylum Cnidaria (Jellyfish and Portuguese Man-of-War)

The cnidarians present as main characteristics a gelatinous body and tentacles that are used to capture food. They undergo a dimorphic life cycle, with a free form of sexual reproduction (the medusa or jellyfish) and a fixed form of asexual reproduction (the polyps). Four important classes cause injuries in humans: Anthozoa (corals and anemones, without the medusa stage), Hydrozoa, Scyphozoa, and Cubozoa (Cubomedusae) [8, 9, 10].

The most severe envenomations are caused by Cubomedusae (especially by the species *Chironex fleckeri* and also by *Chiropsalmus quadrumanus*, *Tamoya haplonema, Carukia barnesi,* and others). Portuguese man-of-war (*Physalia physalis* and *P. utriculus*) also can cause severe envenoming (Figs. 2.8, 2.9, 2.10, 2.11, 2.12a–d, and 2.13) [11–16].

There are hundreds of documented deaths caused by contact with Cubomedusae worldwide (and few caused by Portuguese man-of-war), and most of them are caused by the species *Chironex fleckeri* in the Indo-Pacific region [10]. There are also reports of deaths by jellyfish of order Cubomedusae, *Chiropsalmus quadrumanus,* in the Atlantic Ocean [4, 10]. The species *Chiropsalmus quadrigatus* of the Indo-Pacific region is actually classified as a new species of *Chironex* box jellyfish (*Chironex yamaguchii*).

Despite the simplicity of their body structure, these animals have complex venomous structures. The defense cells called cnidocytes contain nematocysts and other organelles composed of small spicules distal to a spiral structure kept under pressure. The nematocysts fire under changes in environmental pressure and/or osmosis and inoculate venom deeply into the dermis of the humans (Fig. 2.14). Some cnidarians, as the Portuguese man-of-war, contain tentacles

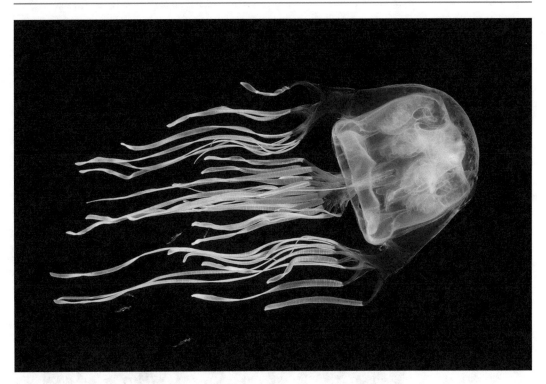

Fig. 2.8 The box jellyfish *Chironex fleckeri* is the most dangerous jellyfish in the world. Dozens of deaths are associated with the envenomation caused by this cnidarians. (Photo: Gary Bell/OceanwideImages.com)

measuring more than 30 m and these have millions of nematocysts that inject micro dosages of venom and cause very serious envenomation. It is important for the emergency care professional to know that many nematocysts initially remain intact in the victim's skin, without discharging its contents [4, 10].

The venom of nematocysts contains various active substances, such as tetramine, 5-hydroxytryptamine, histamine, and serotonin. The main toxic fractions, however, are thermolabile, high molecular weight proteins and peptides capable of altering ionic permeability, thereby causing muscular dysfunction (including cardiac toxicity) [4, 10]. The venom can also cause hemolysis and associated renal failure, which are observed at experimental level but possible in clinical late phases of the envenoming as well [10]. A very important fact that emergency centers should note is that nontoxic proteins can trigger severe allergic processes, including anaphylactic shock [1–5, 10]. The concentration and potency of the venom (and the severity of the

envenoming) vary on a crescent scale from the corals and anemones to the Portuguese man-of-war and some species of jellyfish, such as the box jellyfish [1–5, 10].

Due to the toxic and allergenic properties of the venoms, we can observe direct toxicity and/or allergic phenomena in the envenoming by cnidarians. The toxic action is immediate and the allergic actions are immediate and delayed. The great immediate marker of the envenoming caused by a free cnidarian is the intense pain that appear immediately in the points of contact with the animal. The sensation is similar to a pain caused by a burn, but it should not be called burn as it is the result of an action of toxins.

The lesions have the following characteristic: the tentacles (and occasionally, the body) of the animal are "printed" on the skin, presenting as a crossed linear erythematous papular rash of an urticariform nature. The edematous lines can present horripilation, probably by altering the sympathetic nervous system [1–5]. Within hours, the place of contact may develop vesicles, blisters, and superfi-

Fig. 2.9 *Chiropsalmus quadrumanus*, a box jellyfish found in the Atlantic Ocean. This cnidarian can produce severe lesions in bathers. (Photo: Alvaro Esteves Migotto, São Paulo University)

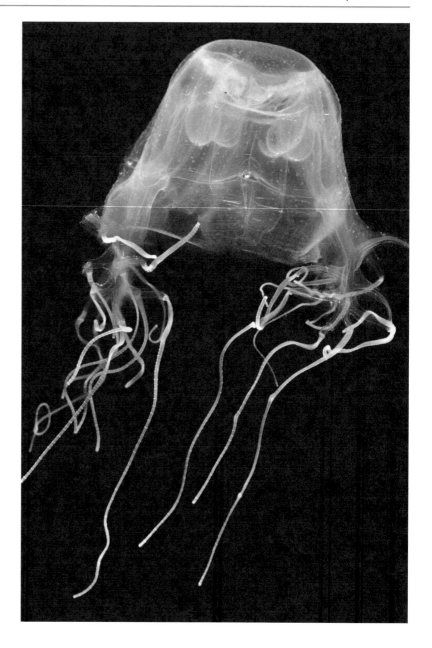

cial necrosis. In mild envenoming, pain decreases in hours, but severe manifestations can cause systemic phenomena such as a general disorganization of nerve activity, heart failure (rare), cardiogenic shock, respiratory failure, hemolysis, and renal abnormalities. Cardiorespiratory manifestations are responsible for deaths in severe cases.

Contact with cnidarians (in sequence) can precipitate emergency allergic reactions such as angioedema and anaphylactic shock, but late reactions are possible: persistent lesions after 48 hours, new lesions at distance, recurrent reactions (four or more), contact dermatitis, or late onset of new lesions. Ingestion of jellyfish is observed in oriental cuisine and is associated with gastrointestinal and skin allergies [17].

If a victim presents a small number of crossed long lines (near 20 cm), there is a probability of severe envenoming with systemic manifestations, and the patient should be monitored for some

Fig. 2.10 The image shows the box jellyfish *Tamoya haplonema* on the sand of a beach in the Southeast region of Brazil. (Photo: Aldo de Aguiar Falleiros, São Vicente, Brazil)

hours, with attention for cardiac arrhythmias and respiratory manifestations. This profile is observed in injuries caused by Cubomedusae and Portuguese man-of-war (Figs. 2.15, 2.16, 2.17, 2.18, 2.19, 2.20 and 2.21).

The most lethal jellyfish probably is the box jellyfish *Chironex fleckeri*. Contact with this Indo-Pacific jellyfish causes great linear marks with a "frost" aspect at the surface of the lesions and an excruciating pain (Fig. 2.22). The venom can turn the cells porous, which causes hyperkalemia and cardiac failure. The less serious sting can be controlled with cold water compresses and baths with vinegar. In serious envenoming, antivenom produced in Australia shall be applied quickly. Until the antivenom can be used, it is very important to maintain the vital signs of the victim [18–20].

The majority of the envenomations by *C. fleckeri* do not cause serious manifestations, but about 64 deaths were registered in Australia since 1883 [18, 19].

The Irukandji jellyfish is a group of small jellyfish capable of provoking serious systemic manifestations in the victims. The marks of contact are very small and sometimes they cannot be visualized. The signs and symptoms caused by the venom are reunited in a syndrome called Irukandji, named by the Australians aborigines. Actually there are four species associated with the syndrome, including the original species *Carukia barnesi*: *Malo kingi*, *Alatina alata*, and *Malo maximus* [18–20]. Small amounts of venom produce catastrophic effects in the liberation of catecholamines acting through sodium channels. The pain of sting is mild, but the patient can present with cold sweating, nausea and vomits, intense muscular pain and cramps, dorsal pain, facial flushing, tachycardia, and arterial hypertension in a period of 5 minutes to 2 hours after

Fig. 2.11 *Carukia barnesi* is one of the jellyfish associated with the Irukandji syndrome in the Indo-Pacific. (Photo: Ilka Straehler-Pohl, Germany)

the contact. Hospitalization is needed and the manifestations can last weeks. Treatment utilizes symptomatic drugs such as antihistamines and anti-hypertensive drugs [19, 20].

However, the marks after the contact with jellyfish are not always long and edematous lines. Small species can have small tentacles and the contact causes only pain and small marks, including round or oval marks on the body (Figs. 2.23 and 2.24). This occurs around the world, but it is particularly noticeable in some South American species, such as *Olindias sambaquiensis* and *Chrysaora lactea*, which are responsible for the majority of the envenomings in Southern Atlantic Ocean (Figs. 2.25 and 2.26) [12, 15, 21]. In this type of envenoming, it is rare to observe severe systemic manifestations and the allergic phenomenon is the great problem to the health professionals.

Other genera and species of jellyfish can cause human injuries around the world: the scyphozoan of the genera *Chrysaora* (Fig. 2.27) and *Pelagia* (Figs. 2.28 and 2.29) are associated with

envenomations in Europe, so as the species *Cyanea capillata* (the lion's mane jellyfish) in cold waters of the Atlantic and Pacific oceans. This species is the largest jellyfish and their tentacles can measure up to 35 m (Fig. 2.30). The *Cyanea capillata* jellyfish is cited in the short story "The Adventure of the Lion's Mane" in the book "The Case-Book of Sherlock Holmes" of Sir Arthur Conan Doyle.

The planulae (larvae) of the scyphomedusae *Linuche unguiculata*, the thimble jellyfish (Figs. 2.31 and 2.32), are associated with the sea-bather's eruption, an intensely pruritic dermatitis manifested by erythematous papules that develop in areas covered with swimsuits [22–24]. The lesions are very characteristic, developing under the swim clothes when the larvae are captured among the fibers of the clothes and fire their cnidocytes (Fig. 2.33). The disease is common in the Caribbean and South of USA.

The injuries caused by jellyfish present with typical lesions that help in the identification of the causative agent by the health professionals, whereas the injuries caused by some hydrozoans, true corals, and anemones show pain or burning symptoms without a typical pattern. The marks are irregular, presenting an erythematous plaque with a rounded/oval shape or papules/vesicles dispersed in the area of contact. These patterns of lesions are observed in divers who had contact with the underwater substrate.

Erythematous, irregular, and painful plaques and papules arise at points of contact with the short tentacles of anemones (Fig. 2.34). On the European coast, especially in the Mediterranean Sea, anemones are a problem. *Anemonia sulcata* and *Actinia equina* are shallow water anemones, which often come into contact with bathers and cause irregular, erythematous, and painful lesions that persist for several days (Figs. 2.35 and 2.36). Systemic manifestations can occur, such as malaise, cramps, and allergic phenomena, which can reach anaphylactic shock, and there are reports of deaths by these cnidarians [25].

The fire corals (*Millepora* sp.) are common in the Caribbean and South America regions. They are in fact hydrozoans that cause severe and extensive envenoming and may lead swimmers and divers to medical emergencies (Figs. 2.37 and 2.38) as other hydrozoans [26]. True corals show minor toxicity but can provoke extensive and deep wounds in bathers (Figs. 2.39, 2.40, and 2.41).

The most common complications of injuries by cnidarians are residual hyperpigmentation, keloids, atrophy of subcutaneous tissue, and, rarely, gangrene [1–5]. Cuts by corals predispose foreign body granulomatous reactions due to the retention of fragments of the exoskeleton of the cnidarians, which is composed of calcium carbonate. Occasionally, corals can provoke acute urticarial lesions. These usually happen in fauna researchers, who touch the animals when they are

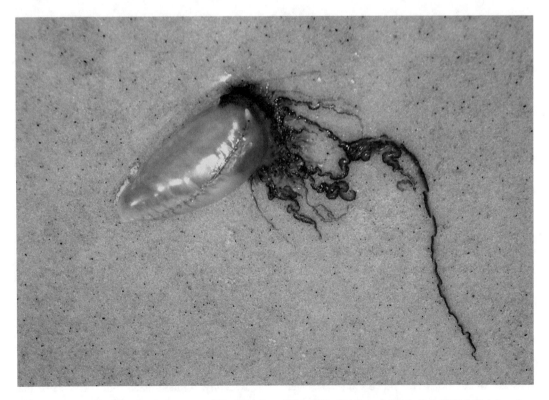

Fig. 2.12 (**a–d**) *Physalia physalis*, the Atlantic Portuguese man-of-war collected in the Southeast Coast of Brazil (Rio de Janeiro State), Northeast Coast (Maranhão State), and North Coast (Pará State) region. (Photos: Ana Maria Mosca de Cerqueira (Rio de Janeiro) and Vidal Haddad Junior)

Fig. 2.12 (continued)

collected. A typical example is the urticarial plaques caused by the snowflake coral *Carijoa riisei* in a professional who comes in contact with the cnidarian while collecting them (Fig. 2.42).

There are few methods to aid in the diagnosis of envenoming by cnidarians: the seabather's eruption can be confirmed by enzyme-linked immunosorbent assay (ELISA) [27]. Histological exams can help identify acute envenoming by the presence of cnidocytes that have semi-penetrated the skin. Some species of cnidarians could be identified by the nematocysts recovered from the human skin by means of adhesive tape method.

Late allergic phenomena can be demonstrated by contact tests.

The treatment of envenomings caused by cnidarians actually is controversy: cold sea water compress or the application of cold packs is a rule. Low temperature promotes analgesia and it can be applied even on the beach [1–5, 12, 15, 16]. Sea water is indicated because the presence of freshwater fires nematocysts fixed in the skin by osmosis. Recently, a randomized trial showed that the use of hot water (45 ° C) for 20–30 minutes would be more effective than cold water [28]. But we think that the nociceptive activity of the venom is

Fig. 2.13 *Physalia utriculus* , the species of Portuguese man-of-war of the Pacific Ocean, has only one long tentacle. (Photo: Juan Pedro Lonza, Iquique, Chile)

influenced by extreme temperatures, not exclusively by hot or cold water, since cold water also has good analgesic effect. This is a fundamental aspect to be considered since the application of cold water has been the clinical indication for years with good results. However, if the application of hot water on the injured site really decreases the painful events, then we must consider that both hot and cold water are effective in producing an analgesic effect. A large-scale evaluation regarding this is in its final evaluation phase, whose data are compatible with this idea [1–5, 12, 15, 16].

When the animal involved is a Cubomedusae (*Chironex fleckeri, Carukia barnesi, Tamoya haplonema,* or *Chiropsalmus quadrumanus*), 5% acetic acid (vinegar) is applied to inactivate nematocysts that are still intact on the skin or in the tentacles not removed. In the envenomation by Portuguese man-of-war (*Physalia physalis)* or scyphozoans, the orientation is not as secure, and in vitro experiments show that the nematocysts of some specimens of Portuguese man-of-war fired when placed in solutions of vinegar or alcohol [29]. Again, certain

experimental studies show that vinegar block entire nematocysts but stimulate those that fired to fully empty their content, but in our clinical experience, application of vinegar proved to be a beneficial measure for any injury caused by cnidarians of South America, including Cubomedusae and Portuguese man-of-war [1–5, 12, 15].

When no treatments could be applied, one should wait for the arrival of a health team; treatments that were certainly popular once, such as use of alcohol, urine, antihistamines, or Coca-Cola®, can aggravate the envenoming. Some patients can present with severe and potentially fatal systemic manifestations, such as arterial hypotension/hypertension, direct cardiotoxicity with cardiac arrhythmias, and secondary pulmonary edema. A patient with these signs and symptoms shall be referred urgently to a hospital. The persistent and intense pain after first aid measures also shall be treated in a hospital (in these cases, an ampoule intramuscular dipyrone seems to be able to control the pain). Cardiac arrhythmias should be treated with intravenous use of verapamil.

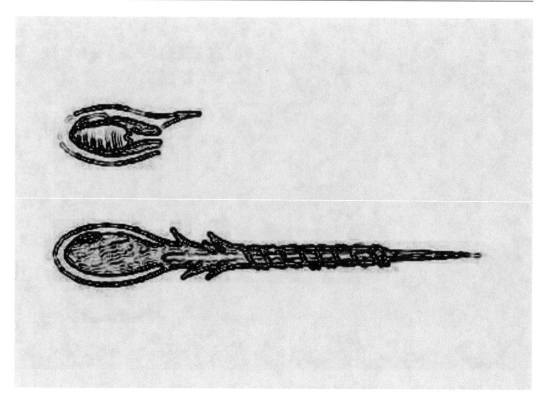

Fig. 2.14 Cnidocytes contain explosive venomous cnidocysts or cnida, which include the nematocysts. These micro apparatus inject venom when the discharge mechanism is activated by osmotic pressure or when the hair-like structure cnidocil is touched (see the unfired nematocyst above in the figure). The needle penetrates into the human dermis. (Photo: Vidal Haddad Junior)

Although sporadic envenomations are common in tropical and semi-tropical marine areas, injuries by cnidarians in large numbers are associated with an increase in animals in certain regions, a fact observed with seasonality. Certain maritime areas used for leisure activities in the Southern region of Brazil register hundreds of thousands of envenomations in the summer season, an unexpected number, when we consider that most contacts are occasional [30–34].

There are two species of jellyfish associated with these injuries: *Olindias sambaquiensis*, a hydrozoan, and *Chrysaora lactea*, a scyphozoan. Both cause envenomation of low to medium severity, causing severe pain, but rare systemic phenomena, which when they occur are most often associated with allergic phenomena [30–34].

The outbreaks of thousands of injuries are caused by one or another species. With this increased number of contacts, some interesting aspects can be observed, such as the involvement of athletes in swimming competitions, creating problems so as to interrupt the competition, and observation of an anaphylactic shock in a child, which is already described as a possible complication of jellyfish envenomation [30–34]. Although there is certainly an increase in jellyfish, other factors contribute to this alarming statistic: the large number of bathers on vacation, appropriate climate for the reproduction of the species mentioned, changes in the environment precipitated by human and natural actions, and others. Interestingly, injuries reports dropped exponentially during the Coronavirus pandemic, reinforcing the importance of massive human presence in the genesis of outbreaks.

Recently, new clinical aspects have been published about the lesions observed in the seabather's eruption, in addition to the classic

Fig. 2.15 Great linear plaque caused by Portuguese man-of-war (*Physalia physalis*). (Photo: Adriana Lucia Mendes, São Paulo)

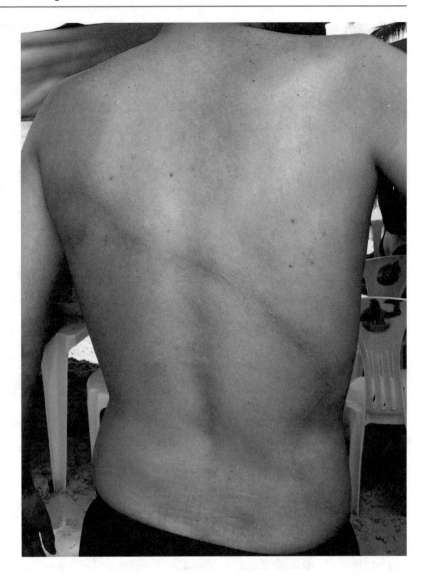

erythematous and itchy papules in areas covered by bathing suits: the presence of Koebner's phenomenon, with reproduction of lesions in areas of trauma due to pruritus, has been described, as well as lesions of larger diameter, which can possibly be caused by adult forms of jellyfish *Linuche unguiculata* and *Linuche aquila*, which are described as the causative agents of seabather's eruption in the Philippines. These findings broaden the range of lesions to be looked for in the suspicion of a manifestation of seabather's eruption [35–38].

Another problem that occurs on the Brazilian coast is the injury caused by the Portuguese man-of-war in the Northeast region, more specifically in the São Luís City, capital of the state of Maranhão. In this place, there are no blooms, and the almost equatorial situation of the region's climate is added to the fact that São Luís is located on an island, which at certain times causes sea currents and winds to bring animals close to the beaches.

In some urban beaches, such as Calhau Beach, envenomations are common, reaching about 300 occurrences per weekend, with a large influx of Portuguese man-of-war of various diameters to the beaches. This makes the bathers in the water to come in contact with these animals, and people (especially children) touch them out of curiosity

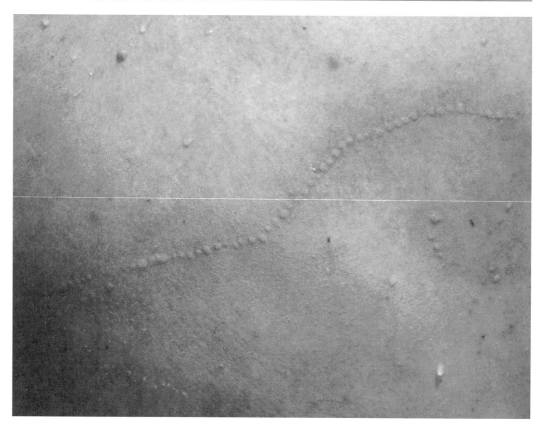

Fig. 2.16 Detail of the lesions showing the "batteries" of cnidocytes in the segments of the tentacles marked in the skin. (Photo: Adriana Lucia Mendes, São Paulo)

due to the bright colors they have. Envenomation by Portuguese man-of-war are more important than by other cnidarians (with the exception of Cubozoans) and deserve clarification campaigns and provision of first aid in the areas [39–40].

Injuries caused by cnidarians commonly are easy to identify due to the intense pain that arises immediately after the contact with the victim still in the water. Additionally, a classic pattern of erythematous and edematous crossed lines is observed on the skin. It is very important to observe that clearly there is a characteristic aspect in the clinical manifestations caused by cnidarians in the Atlantic Ocean in South, North, and Central America: few, long, and crisscrossed lines suggest contact with Cubomedusae and Portuguese man-of-war (severe envenomation, excruciating pain, and systemic phenomena). This kind of lesions and visualization of a blue or purpura float denounce the Portuguese man-of-war. However, rounded or oval or diffuse skin lesions, sometimes with impression of small tentacles, are marks caused by the body of small medusae. There are no systemic phenomena (only when there is allergy) and the lesions are suggestive of those caused by *Olindias sambaquiensis* and *Chrysaora lactea* and other small hydrozoans and scyphozoans [15].

Fig. 2.17 Purpuric plaques after envenomation by *Physalia physalis*. (Photo: José Yamin Risk)

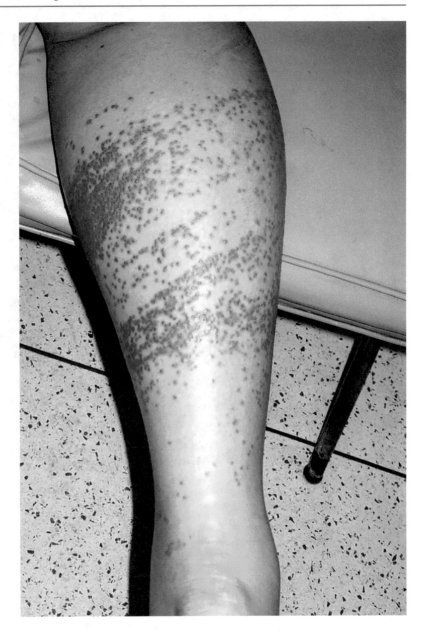

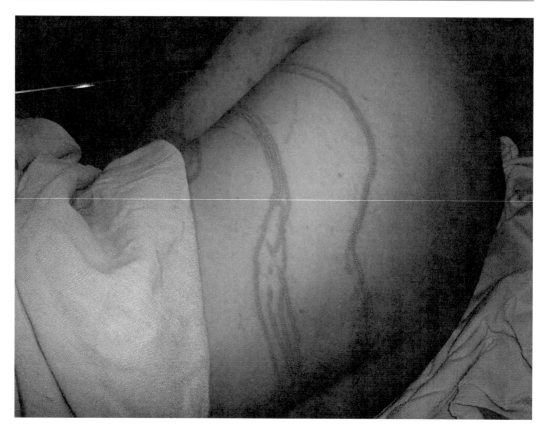

Fig. 2.18 Long linear plaques caused by *Physalia physalis*. These lesions are compatible with envenomations caused by Portuguese man-of-war and box jellyfish. (Photo: Vidal Haddad Junior)

Fig. 2.19 This patient presented the classic lesions by *Physalia physalis* after contact with a specimen in a beach of Southeast region of Brazil. (Photo: Fernando Croitor, Brasília)

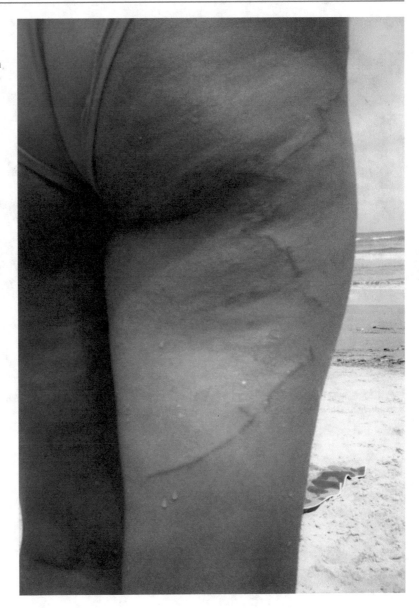

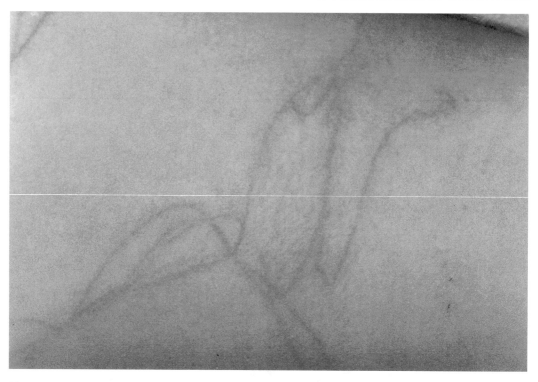

Fig. 2.20 Linear plaques with the classical pattern in "crossed lines" suggestive of envenomations by *Physalia* sp. and box jellyfish. The image was made 1 hour after the injury. (Photo: Vidal Haddad Junior)

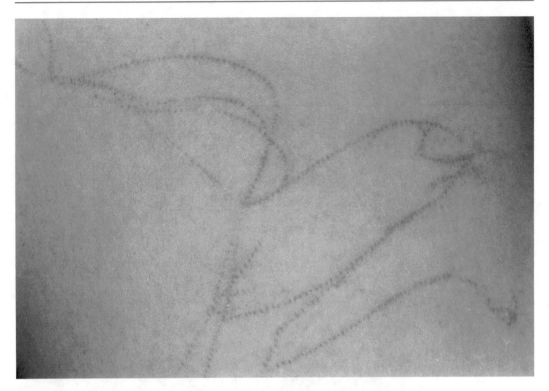

Fig. 2.21 The same area 24 hours after the contact. Superficial skin necrosis is observed. (Photo: Vidal Haddad Junior)

Fig. 2.22 The patient seen in the image is a surfer who had contact with *Chironex fleckeri* in the Australian coast. He did not present severe systemic manifestations, despite the extension of the plaques. (Photo: Guto Amorim (www.waves. com.br))

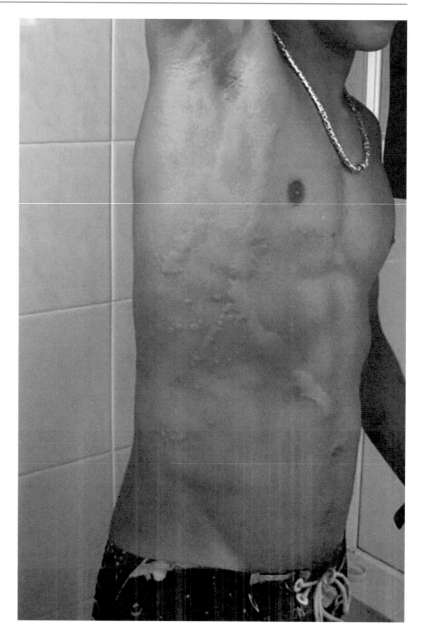

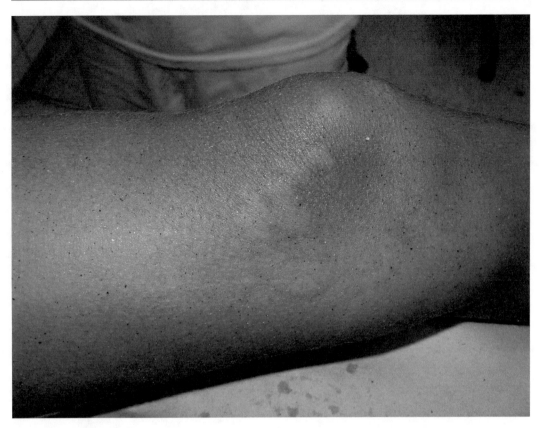

Fig. 2.23 Round edematous marks associated with intense pain in a patient who suffered an envenomation by *Olindias sambaquiensis* in the South region of Brazil. (Photo: Vidal Haddad Junior)

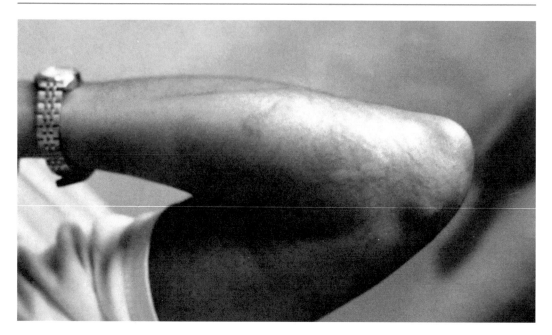

Fig. 2.24 Irregular edematous plaques after contact with *Chrysaora lactea*. (Photo: Vidal Haddad Junior)

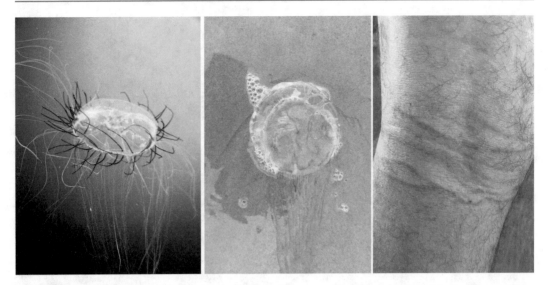

Fig. 2.25 The hydrozoan *Olindias sambaquiensis* is responsible for most of envenomations in Southeast and South regions of Brazil and Atlantic Coast of Uruguay and Argentina. (Photo: Fábio Lang da Silveira, São Paulo University). A specimen of *Olindias sambaquiensis* on the sand and an envenomation caused in a bather. (Photos: Maurício Azevedo O. Costa)

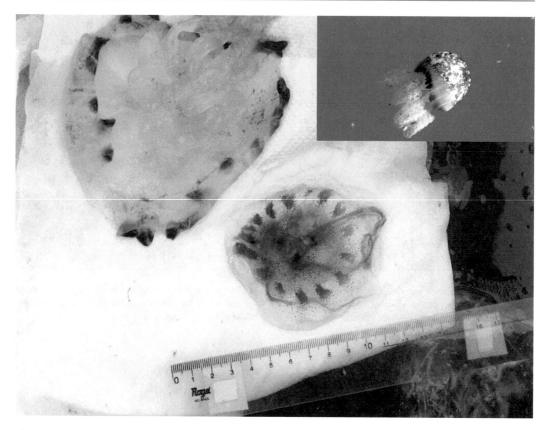

Fig. 2.26 The image shows two scyphozoan jellyfish: *Lychnorhiza lucerna* (inset) and *Chrysaora lactea* (main figure). The second is associated with a great number of envenomations occurred during holiday periods in the South region of Brazil. (Photo: Vidal Haddad Junior)

Fig. 2.27 A large specimen of *Chrysaora* sp. (Photo: Vidal Haddad Junior)

Fig. 2.28 The most common species of jellyfish in East Atlantic is the scyphozoan *Pelagia noctiluca* associated with human injuries in the European Coast and associated islands. (Photo: João Pedro Barreiros, Azores Islands)

Fig. 2.29 Edematous and erythematous plaques associated with intense pain after envenomation caused by *Pelagia noctiluca* in the Azores Islands. (Photo: João Pedro Barreiros, Azores Islands)

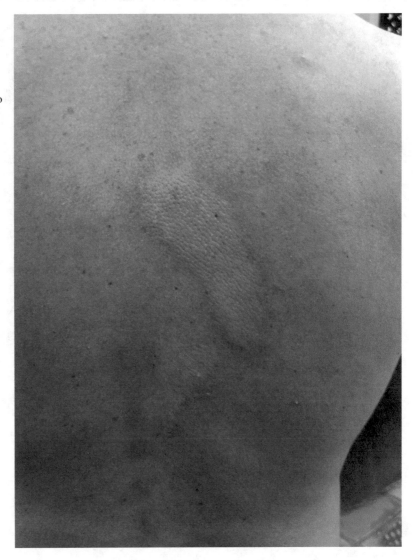

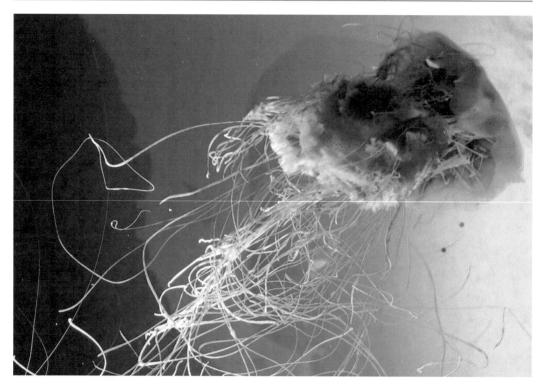

Fig. 2.30 *Cyanea capillata*, the lion's mane jellyfish. This large species is capable of causing envenomations in human beings. (Photo: Vidal Haddad Junior)

Fig. 2.31 The thimble jellyfish (*Linuche unguiculata*) is the agent of the seabather's eruption, common in the West Atlantic Coast. (Photo: Fábio Lang da Silveira, São Paulo University)

Fig. 2.32 The planulae larvae of the thimble jellyfish is responsible for dermatitis under the bathing suits, by becoming trapped in the tissue and firing their cnidocyte. (Photo: Fábio Lang da Silveira, São Paulo University)

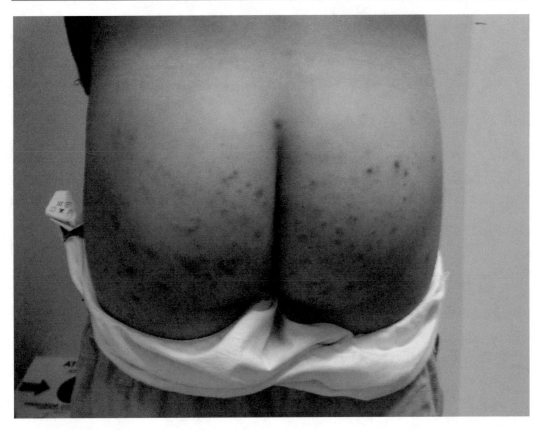

Fig. 2.33 The erythematous and pruritic papules under the sea clothes are typical of the seabather's eruption. This patient was observed in a beach of the South region of Brazil. (Photo: André Luiz Rossetto, UNIVALI, Brazil)

Fig. 2.34 Anemone showing the short tentacles that can cause injuries manifested by little characteristics marks on human skin. (Photo: Vidal Haddad Junior)

Fig. 2.35 *Actinia equina* or snakelocks anemone, a common anemone in the Mediterranean Sea, associated with human injuries. (Photo: Vidal Haddad Junior)

Fig. 2.36 *Anemonia sulcata* is a common causer of envenomation in the Mediterranean region. (Photo: Vidal Haddad Junior)

Fig. 2.37 The *Millepora* genus (fire corals) are hydrozoans that cause envenomations mainly in divers in Western Atlantic Ocean. (Photo: Fábio Lang da Silveira, São Paulo University)

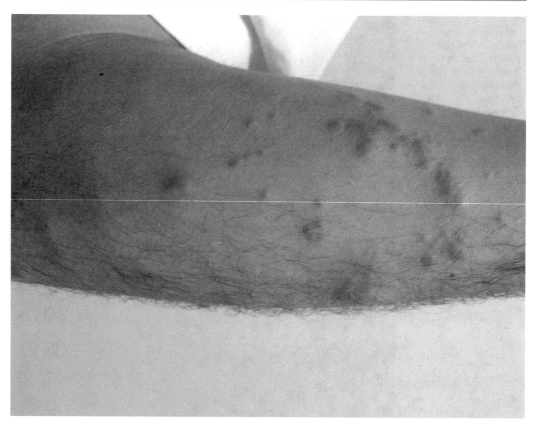

Fig. 2.38 The irregular and finger-like plaques and papules caused by the fire corals are distinctive of the classic lesions caused by the jellyfish. (Photo: Vidal Haddad Junior)

Fig. 2.39 A brain coral contains calcareous skeleton and live polyps. It can provoke excoriations and envenomation. (Photo: Álvaro Esteves Migotto, in the site of CeBiMar/São Paulo University)

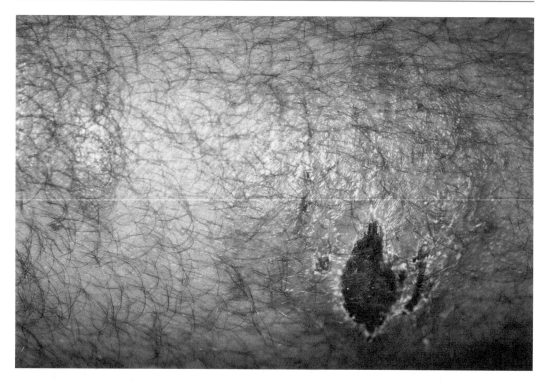

Fig. 2.40 Contact with calcareous skeleton of the corals provokes cuts and abrasions with possibility of severe infections. (Photo: Vidal Haddad Junior)

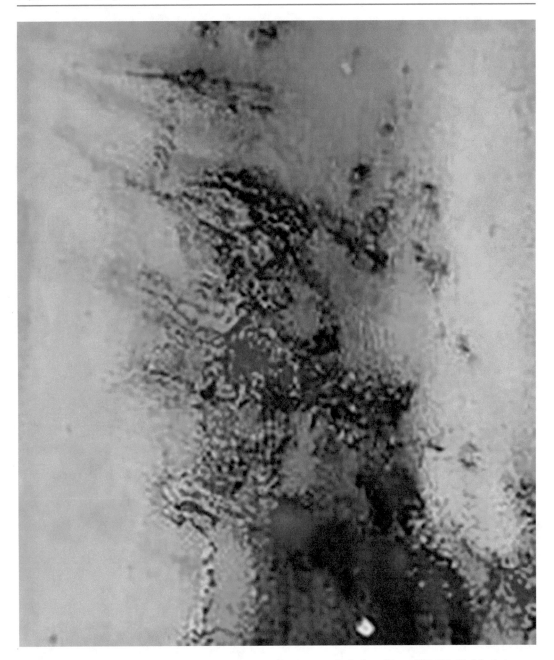

Fig. 2.41 Extensive ulcerations provoked by coral cuts in the dorsum of a bather. (Photo: Vidal Haddad Junior)

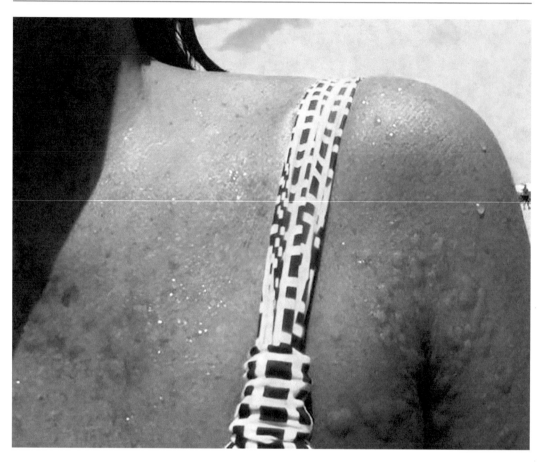

Fig. 2.42 Urticariform plaques immediately after the contact with the octocoral *Carijoa riisei*. (Photo: Fernanda M D Amaral, Universidade Federal Rural de Pernambuco)

Box 2.1 Cnidarian – Mild/Moderate Envenomation

A 13-year-old male, upon entering the sea with water on the chest level, felt something touching his body and almost immediately expressed intense pain. When he came out of the sea, he noticed an erythematous and edematous round plaque on the abdomen with associated small linear plaques (see image). The pain increased in intensity and the patient was instructed to apply urine on the injuries, not getting better and seeking hospital care. On examination, the patient was agitated, with normal arterial tension. The pain was very strong and he was told to use cold seawater compresses on the compromised areas and take vinegar baths each 30 minutes. About 15 minutes after the start of therapy, the pain was tolerable and the patient was calmer. He was discharged after the pain subsided, 2 hours after being admitted to the hospital.

Comments: the patient suffered mild envenomation without systemic involvement and showed systemic changes due to pain caused by the toxins. The injury probably was not associated with most severe injuries caused by Portuguese man-of-war or cubomedusae (see the text), and more likely the pain was caused by the species of scyphomedusae or hydromedusae, common in Brazilian coast. In this area (South of São Paulo State), the most common jellyfish associated with this kind of injury is *Chrysaora lactea* (see image). The first aid treatment was effective.

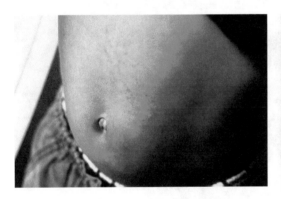

Box 2.2 Cnidarian – Severe Envenoming

The patient was male, 22-year-old, and surfer. He was practicing the sport in deep waters of Aruba, in the Caribbean, and while swimming carrying the board, he was touched by something that he did not see, but due to previous accidents, he knew that it was a jellyfish. Immediately, he felt excruciating pain in the chest, and once out of the water, he did look for lifeguards on the beach. The region showed 15-cm long, linear, erythematous, and edematous plaques (see image). Still on the beach, he presented with dyspnea and "palpitations," which he could not describe in detail. He was taken to the city hospital, and he showed significant tachycardia with occasional extrasystoles, snoring, and congested lungs. The pain, very strong, decreased with the use of cold packs. The patient was treated symptomatically and had gradual improvement. He remained at the emergency room for about 12 hours, before being discharged.

Comments: the long intersecting and linear plaques in small number are very suggestive of envenomation by box jellyfish and Portuguese man-of-war (see image). These injuries are severe, with real possibility of emergencies due to systemic phenomena, including heart conditions and lung-related phenomena. Worldwide, the cubomedusas are responsible for most of the deaths associated with cnidarians and they are suspected of causing this envenomation, because these colored Portuguese man-of-war spend most of the time floating at the water's surface and therefore are encountered by the victims.

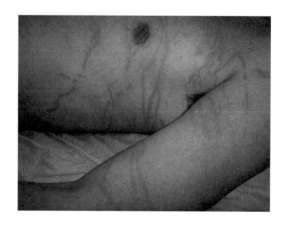

Phylum Echinodermata (Sea Urchins, Starfish, and Sea Cucumbers)

The echinoderms are marine animals that contain different body formats. The sea urchins (Echinoidea class) show a round body covered by hollow traumatogenic spines composed of calcium carbonate, while the starfish (Asteroidea class) have a star format and the sea cucumbers (Holothuroidea Class) have cylinder-shaped body (Figs. 2.43, 2.44, 2.45, and 2.46). The "crown-of-thorns" starfish and sea urchins can contain venoms. The venom of sea urchins has hypotensive, hemolytic, neurotoxic, and cardiotoxic effects, which are due to the toxins present in the pedicellaria, that is, small tentacles that are located among the spikes [4, 41]. Sea cucumbers

Fig. 2.43 Atlantic Black sea urchin (*Echinometra lucunter*). This species is responsible for the majority of human injuries in Atlantic Ocean. (Photo: Vidal Haddad Junior)

Fig. 2.44 The crown-of-thorns starfish is venomous and can provoke painful punctures in humans. (Photo: Vidal Haddad Junior)

(*Holothuria*) produce holothurin-A, a toxin that causes irritation to the skin and mucous membranes. As mechanisms of defense, they discharge sticky internal structures (Cuvierian tubules) from the anus to immobilize predators or liberate a jet of toxin-containing liquid that can cause severe dermatitis or ocular inflammation in humans and may cause blindness (Fig. 2.47).

Although the majority of the injuries caused by sea urchins is traumatic, as observed in the injuries caused by the black sea urchin *Echinometra lucunter* (Fig. 2.48), some species can be venomous, such as the *Diadema* sp. (Fig. 2.49). Trauma is the main problem caused due to the penetration of and the spines of sea urchins, and the spines can be visualized as small black, white, purple, or green spots on the skin (Fig. 2.50). It is possible to extract fragments up to 3,0 cm from the site, but most of them are small pieces and sometimes there are only pig-

ments at the site of entry of the spine (Fig. 2.51). When the injury is mainly traumatic, the pain is moderate and it only occurs after compression [41–43]. Injuries by venomous echinoderms cause severe skin inflammation manifested by erythema, edema, papules, vesicles, and occasional necrosis [2–5]. The preferential affected regions are the plantar areas (Figs. 2.52, 2.53 and 2.54). The spines can carry secondary infections, including tetanus. Most of the spines are spontaneously eliminated, but their permanent presence may provoke nodules with erythematous and verrucous surfaces (foreign body granuloma) difficult to resolve (Figs. 2.55 and 2.56) [41–43].

The bathers are the major victims of this type of injury, but it is common to see scuba divers presenting spicules or late nodules. The bathers present spines in the feet and suffer injuries when walking in shallow waters and small lakes between the stones of the beaches.

Fig. 2.45 and 2.46 The sea cucumbers are poisonous animals that when devoured can provoke the death of humans. (Photo: Vidal Haddad Junior)

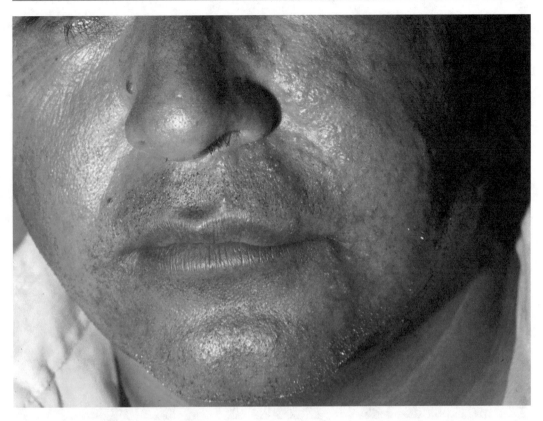

Fig. 2.47 Erythema and edema provoked in the face of a patient caused by a jet of toxins discharged by a sea cucumber. (Photo: Juan Pedro Lonza, Iquique, Chile)

The spines of sea urchins are removed through superficial scarification with a hypodermic needle of large caliber – the same needle used for the withdrawal of spines after local anesthesia (Fig. 2.57). The fragments are brittle and can be difficult to remove, but it is fundamental to extract the large fragments to decrease the possibility of formation of granulomas. Many fragments are expelled by a local inflammatory reaction.

Injuries caused by sea urchins are a common and extremely troublesome problem in some tropical areas. These animals are grouped in lagoons formed by the tides or on rocks on the beaches or underwater, victimizing bathers and divers. Withdrawal is difficult and requires time, and this is not possible in emergency rooms if the conditions are severe and the number of patients are more. We must consider that when there are sea urchins and bathers in the same areas, trauma will happen, due to carelessness or lack of knowledge. The partial but feasible solution is to map the areas where the animals live and to prepare information campaigns, with posters and explanatory leaflets, seeking to alert the population to the "hot spots" of injuries. Prevention is the only safe way to decrease the number of casualties in emergency rooms.

All the venomous animals of this phylum have thermolabile venom, encouraging the use of immersion of the affected site in hot water, around 50 °C, for 30–90 minutes, especially if there is spontaneous pain. This measure can be useful in injuries caused by venomous sea urchins and the starfish "crown of thorns" [10]. The envenomation caused by the irritating secretions of sea cucumbers should be washed thoroughly and any secretion need be removed from the contact sites. When there is ingestion and poisoning, gastric lavage and symptomatic treatment are important.

Fig. 2.48 The species *Echinometra lucunter* causes 50% of the injuries in humans in West Atlantic, but the effects of the spines in the skin are traumatic, once there is no inoculation of venom. (Photo: Vidal Haddad Junior)

Fig. 2.49 *Diadema* sp. is a venomous species of sea urchin and it can causes severe envenomation in human beings. (Photo: Vidal Haddad Junior)

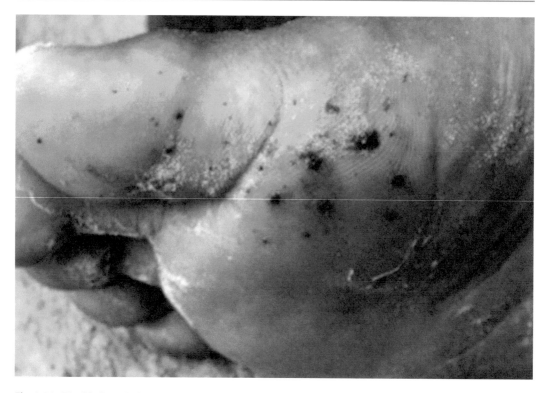

Fig. 2.50 The black marks in the skin are fragments of spines difficult to extract. (Photo: Vidal Haddad Junior)

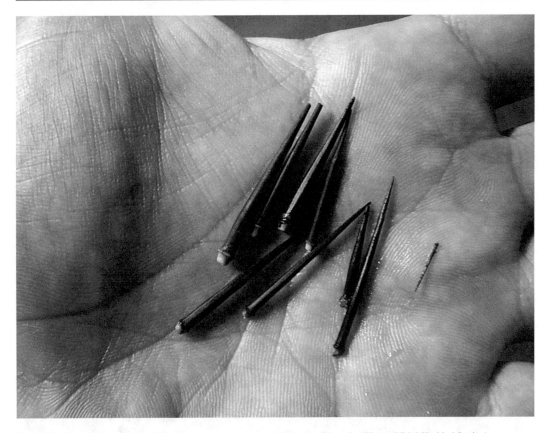

Fig. 2.51 The spines of sea urchins can measure some centimeters of length. (Photo: Vidal Haddad Junior)

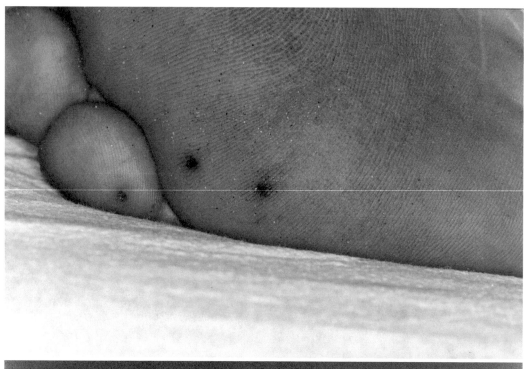

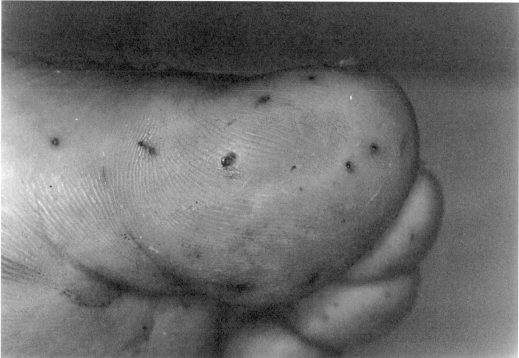

Figs. 2.52, 2.53, and 2.54 The images show typical injuries caused by black sea urchins in bathers. The soles are the main compromised place and the recent lesions do not present inflammation, once there are no envenom-ations. The possible later inflammation is associated with secondary infections and/or foreign bodies present in the skin. (Photos: Vidal Haddad Junior)

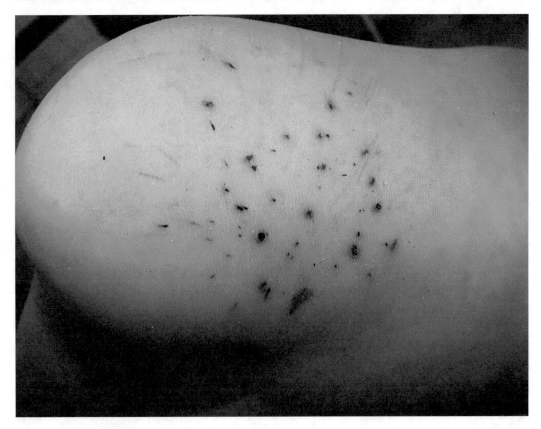

Figs. 2.52, 2.53, and 2.54 (continued)

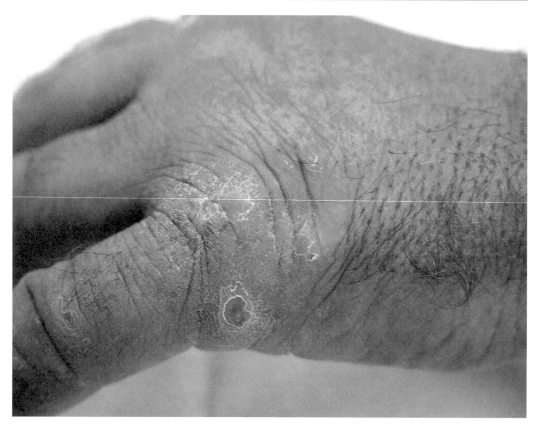

Fig. 2.55 Hyperkeratotic nodules in the point of penetration of sea urchins spines in a diver. The nodules are formed by a foreign body granulomatous reaction. (Photo: Vidal Haddad Junior)

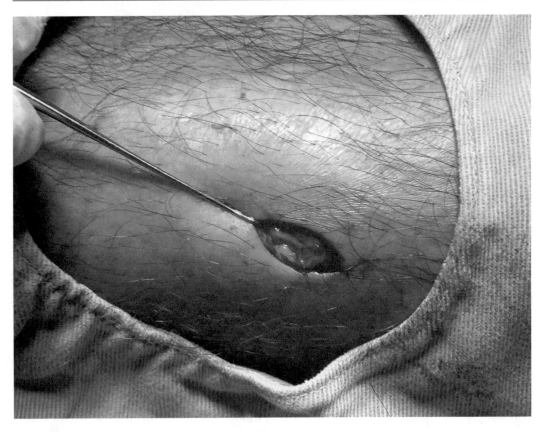

Fig. 2.56 The extraction of the fragments responsible for the foreign body granulomas only is possible after surgical procedures. (Photo: Vidal Haddad Junior)

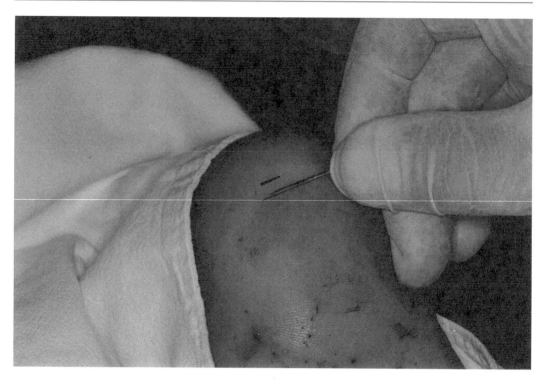

Fig. 2.57 The acute penetration of the spines needs hypodermic needle of large caliber for the extraction. (Photo: Vidal Haddad Junior)

Box 2.3 Sea Urchin

A 17-year-old was walking between lagoons formed by rock formations between two beaches and, due to murky water after rains, he stepped into a colony of black sea urchins (Fig. 2.62), which he observed after the accident. The pain on palpation was important and the right sole showed a dozen of blackened points (Fig. 2.63). He was instructed to remove the spines embedded on the skin. He used a thick needle and withdrew two larger spines, but stopped the process due to pain, resolving to see a doctor. In the hospital, five spines were extracted, the remainder being "too small" for removal. The orientation was to observe the remaining spikes because the local inflamma-tion would make them expelled. Two months after the injury, the remaining spikes were fully eliminated, but at one point, an erythematous, hyperqueratosic, and painful nodule formed, which showed a small area hardened to deep palpation. He was referred to surgery, and a last spine of about 0.2 cm diameter was extracted, resulting in the complete cure of the lesions.

Comments: injuries caused by sea urchins may or may not provoke envenomation. Traumatic accidents, however, require the with-drawal of spines from the penetration points because they cause serious bacterial infections and are responsible for the formation of foreign body granulomas, which are only cured by sur-gical procedures.

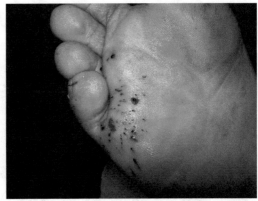

Phylum Annelida (Leeches and Polychetas)

Leeches are worms that are included in the subclass Hirudinea. They are large worms, which can measure up to 15 cm, and are widely distributed around the world. Leeches are founded in freshwater, marine, and even arboreal environments. These worms are hematophagous presenting oral and caudal suckers and jaws with sharp teeth to attach to the victims and feeding (Fig. 2.58). A leech can feed up to ten times its weight of blood, but it does not cause major problems to the victims. The species *Hirudo medicinalis* had widespread therapeutic use in antiquity to promote bleeding of patients. The leeches contain an anticoagulant hirudin in their saliva, which prevents blood from clotting in their gut. Allergic processes and infections may also occur at the site where the worms get attached [2–5].

The marine worms (especially the marine brush worms or fire worms) have chitinous jaws with teeth and/or irritating body bristles (Figs. 2.59, 2.60, and 2.61). They can inflict painful bites in humans, and the penetration of the bristles can provoke cutaneous edema, papules, itch, pain, and skin necrosis (Fig. 2.62).

The place bitten by a marine worm must be repeatedly washed with clean water. Topical antibiotics are useful to prevent bacterial infections. The bristles in the skin must be explored with tweezers and adhesive tape so as to remove them. Injuries by marine worms are not so common with exception of mussels catchers, who are often in contact with these animals. Other potential victims are marine wildlife researchers and divers [2–5].

The tentative of mechanical removal of leeches can cause trauma. Application of alcohol or the heat of a flame seems to be the best ways to remove the worm.

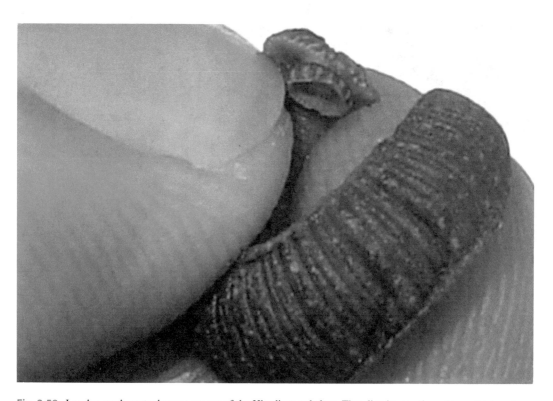

Fig. 2.58 Leeches are hematophagous worms of the Hirudinea subclass. They live in aquatic and terrestrial environments and can prey humans. Note the anterior sucker utilized for feeding. (Photo: Vidal Haddad Junior)

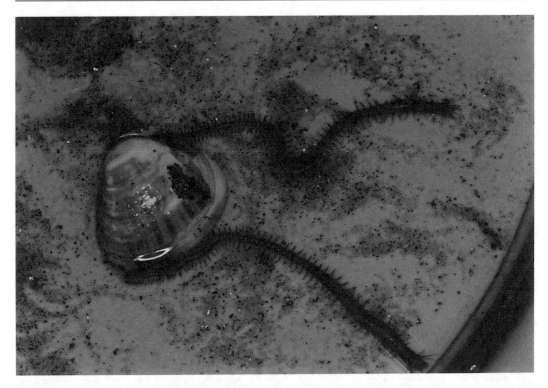

Fig. 2.59 Polychaetas are marine worms that can present bristles or not. The species without bristles can bite with their chitinous jaws. (Photo: Vidal Haddad Junior)

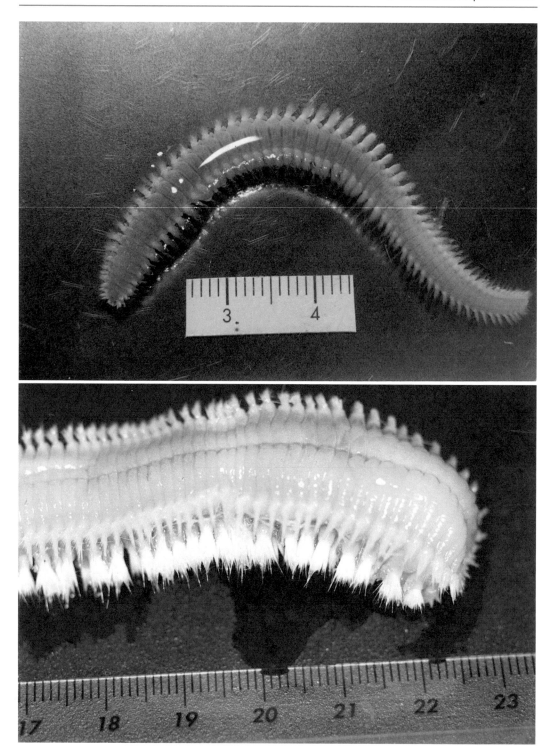

Figs. 2.60 and 2.61 The fire worms or bristle worms present irritant bristles that can cause severe dermatitis in human skin. This type of injury is rare and it is observed in mussel´s catchers. (Photo: Vidal Haddad Junior)

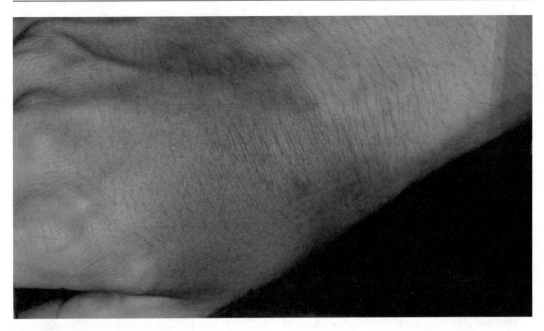

Fig. 2.62 This intense dermatitis manifested by erythema and edema was provoked by contact with bristle worms collected in a Zoology researcher. (Photo: Vidal Haddad Junior)

Phylum Mollusca (Octopuses and *Conus* Snails)

The mollusks are organisms with a shell to protect their soft body, but not all have shells. The shells can be either single or double. Some mollusks are venomous or poisonous.

The *Conus* genus (Gastropoda) causes serious envenomations by injecting venom by a harpoon-like radula that is ejected from the distal end of the proboscis (Figs. 2.63 and 2.64). The venom of *Conus* species is composed of conotoxins, which are low molecular weight neurotoxins. The venom has two different effects – the first effect is the "lightning-strike" effect, which causes immediate immobilization, and the second effect is achieved more slowly, that is, total inhibition of neuromuscular transmission [1–5]. *Conus* that feeds on worms and mollusks causes mild envenomation in humans, but accidents from piscivorous *Conus* can be fatal. During the daytime, they remain inactive under rocks, fragments of shell, and coral. Once collected, they must be handled with care and should not be touched at their opening. One of the most common species found in Brazil is *Conus regius* Gmelin, 1791, which actively feeds on marine worms and is capable of provoking envenomation [44–45].

Region where several specimens of *Conus* are found called the "point of *Conus*". A typical accident from a *Conus* shell initially causes a mild local pain that evolves to progressive muscle paralysis in about 1 hour without other local signs or symptoms. A history of contact with mollusks associated with intense muscular weakness raises a suspicion for this accident. In later phases, the patient can develop palpebral ptosis, blurred vision, speech and deglutition difficulty, unconsciousness, and dyspnea with possible evolution to respiratory arrest which can be fatal and occur between 40 minutes and 5 hours after the sting [2–5, 10]. The number of human fatal envenomations is not known, being estimated at about 50 deaths [44–45].

Among the most dangerous shells are the Indo Pacific species *Conus geographus* (associated with the great majority of the reported deaths), *Conus textile* (also associated with deaths), *Conus magus*, *Conus tulipa*, and *Conus striatus* (Fig. 2.65). There are three Atlantic Ocean species of large diameter: *Conus regius* (Fig. 2.66), *Conus centurio,* and *Conus ermineus* (Fig. 2.67), the second being known as piscivorous [4, 44, 45]. The most venomous species of *Conus* feed on fish.

Cephalopod mollusks are marine animals and include squids, octopuses, cuttlefish, and nautiluses. Octopuses have a horny "beak" that they use to capture prey and for defense mechanisms. In times of danger, they can propel quickly like a jet through water in the opposite direction of perceived threats and they have the ability to eject clouds of dark ink with which they confuse predators. The "beak" can inflict lacerations to victims (especially fishermen and divers), and various species can inject venom present in their salivary glands. The venom contains digestive enzymes and proteinaceous neurotoxins to immobilize prey.

There are four species of blue-ringed octopus in Indo-Pacific region: *Hapalochlaena maculosa* (the Greater blue-ring octopus), *Hapalochlaena lunulata* (the Southern blue-ring octopus), *Hapalochlaena fasciata* (the Blue-lined blue-ring octopus), and *Hapalochlaena nierstraszi*. They are small animals measuring 5–10 cm that show iridescent blue rings when irritated. This feature of displaying vivid and bright colors is called aposematism, by which means these species "warn" their potential predators about the risk they offer.

All of them inoculate maculotoxin and neutransmitters from their salivary glands through one horny beak (it was recently demonstrated that maculotoxin is identical to tetrodotoxin). Tetrodotoxin is a potent neurotoxin that blocks axonal sodium channels and provokes muscular paralysis similar to that observed in envenomations with *Conus* shells, occasionally causing human deaths by respiratory arrest (Fig. 2.68).

Even the common octopus (*Octopus* sp.) has recently been described as a venomous animal (Fig. 2.69). Common octopuses can inject a neurotoxic glycoprotein called cephalotoxin [46]. There is a report of an envenomation caused by a common octopus, which presented generalized neurotoxicity manifested by paresthesias (including perioral), malaise, dizziness, diarrhea, and muscle weakness without paralysis after consumption of raw meat octopus (Japanese cuisine) by a young woman [46]. An injury caused by the "beak" of an octopus in a patient's hand provoked an area of induration and erythema about 8.0 cm which persisted for weeks (Fig. 2.70) [47]. The suckers in the tentacles of octopuses can cause traumatic purpura by the strong suction (Fig. 2.71) [48].

The diagnosis of the envenoming by *Conus* snails depends on the clinical history and the signs and symptoms of the patient. The severity of the disease manifestation depends on the complete occurrence of symptoms, and the risks also depend on the geographical area of contact (the time until the patient is taken care of also influences the prognosis) [2–5, 44–45].

As the manifestations of the envenomations caused by the blue ring octopus and the *Conus* snails are neurologic, including paralysis, it is important to remove occasional tissues of the mollusks in the area of the bite/sting and perform local asepsis. If the patient develops neuromuscular paralysis, then immediately introduce artificial respiration and other measures to control respiratory failure, which is the only effective means of treatment.

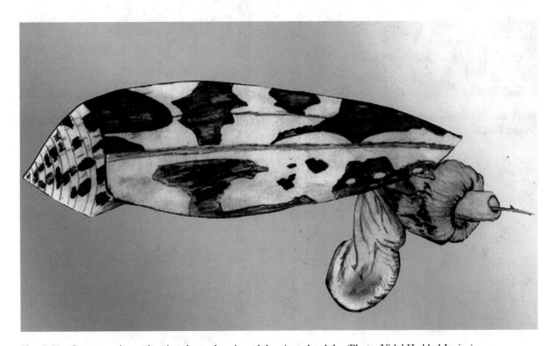

Fig. 2.63 *Conus* specimen showing the proboscis and the ejected radula. (Photo: Vidal Haddad Junior)

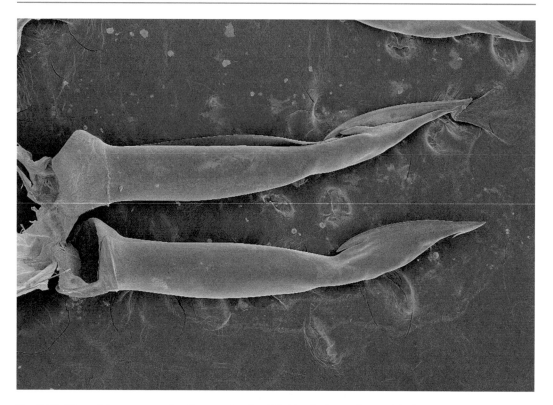

Fig. 2.64 The radula apparatus of a *Conus regius* (modified in the form of several barbed and hollow harpoons). (Photo: Luis Ricardo L. Simone, São Paulo University)

Fig. 2.65 Left to right: *Conus aulicus*, *Conus striatus*, *Conus geographus* (two specimens), *Conus tulipa*, *Conus omaria*, *Conus marmoreus*, *Conus textile*, and two exemplars of *Conus magus*. (Photo: Vidal Haddad Junior)

Fig. 2.66 A live *Conus regius* specimen. This mollusc is a very common species found from the Caribbean Sea to São Paulo coast, Brazil. (Photo: Vidal Haddad Junior)

Fig. 2.67 *Conus centurio* and *Conus ermineus*, two large *Conus* species of the Atlantic Ocean. (Photo: Vidal Haddad Junior)

Fig. 2.68 *Hapalochlaena lunulata* is the southern blue-ring octopus. This small mollusk can present with maculotoxin and produce severe envenomation in humans, including respiratory arrest and death. (Photo: Vidal Haddad Junior)

Fig. 2.69 *Octopus vulgaris* is the most common species of octopus and can present cephalotoxin, a poor studied neurotoxin associated with muscular paralysis. (Photo: Cláudia Alves de Magalhães, Brazilian Ministry of Science, Technology and Innovation)

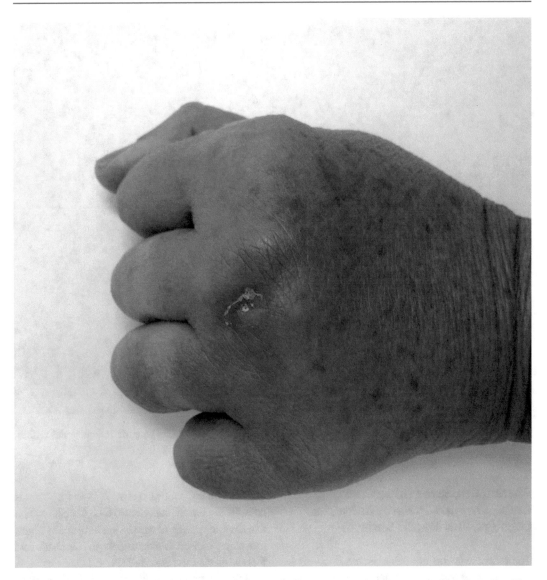

Fig. 2.70 A plaque of chronic evolution with infiltration was formed in the dorsum of the left hand of this patient. Note the central ulceration caused by the beak of the octopus. (Photo: Cláudia Alves de Magalhães, Brazilian Ministry of Science, Technology and Innovation)

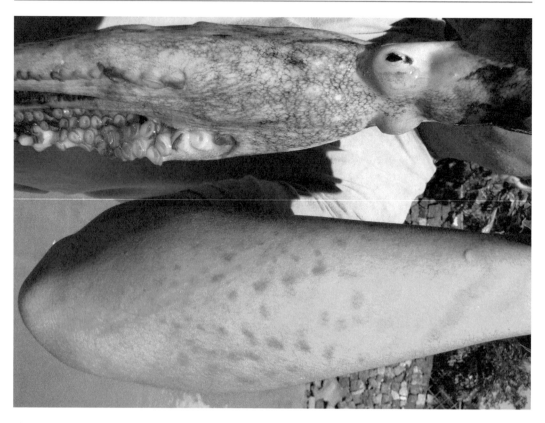

Fig. 2.71 The strong suction of the suckers of the arms of octopuses can provoke purpuric lesions in the victims. (Photo: Rafael Augusto Gregati, zoologist)

Phylum Crustacea (Blue Crabs, Crabs, Shrimps, Prawns, Barnacles, Lobsters, and Mantis Shrimp)

The crustaceans do not cause envenomations by inoculation of toxins. *Speleonectes tulumensis* is the only species recently described and without clinical importance that uses its venom to hunt its preys. However, it can provoke serious poisonings after ingestion of its meat and precipitate severe allergic reactions.

Traumatic injuries are the rule, mainly lacerated wounds caused by its claws. The injuries are not severe and rarely provoke great lacerations or intense bleeding. Crabs and shrimps can provoke this type of lesion (Figs. 2.72, 2.73, and 2.74), but the fishermen fear especially the mantis shrimp, a large and aggressive crustacean (up to 30 cm), whose sharp claws can cause serious injuries to their hands; hence, mantis shrimps are known as *thumb splitter*

in parts of the Caribbean (Figs. 2.75 and 2.76). It is possible to observe cuts caused by barnacles, which live fixed to rocks and woods and have sharp edges, provoking incise wounds mainly in the hands and foot of the victims (Fig. 2.77) [4, 49].

Allergic processes linked to ingestion of or contacts with crustaceans are not rare: all the manifestations of the allergy can occur, such as contact dermatitis, urticaria (Fig. 2.78), and even anaphylactic reactions. Anaphylaxis is a severe complication that is probably associated with cross reactions with tropomyosins present in all crustaceans and astaxanthin, the pigment that gives reddish color to shrimp, crabs, lobsters (crustaceans), salmons, and flamingos [2–5]. Contact dermatitis is manifested by an acute eczematous process. An irritant contact dermatitis is observed in the hands of shrimp and prawn cleaners (Fig. 2.79).

Treatments for injuries caused by crustaceans are as follows: intense washing of the wound,

Fig. 2.72 A crab with the claws in defense position can provoke lacerated wounds in a victim. (Photo: Vidal Haddad Junior)

tetanus vaccination, and use of topical or systemic antibiotics if infection occurs. The use of antihistamines and topical corticosteroids controls contact dermatitis, but anaphylaxis needs fast assessment of the airway and fast transport to a hospital. The dosing of epinephrine for the acute treatment of anaphylaxis is 0.01 mg/kg and a maximum of 0.5 mg can be administered intramuscularly every 5–20 minutes as necessary. Antihistamines and corticosteroids can be useful, but these drugs never will be a substitute for epinephrine. Antihistamine is important for skin lesions and corticosteroids probably decrease the risks of new allergic processes.

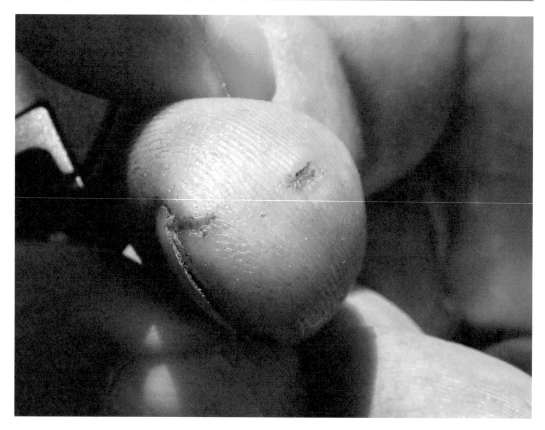

Fig. 2.73 The typical lesion caused by the claws of a crab shows two lacerated wounds caused by the claws. (Photo: Vidal Haddad Junior)

Fig. 2.74 The rostrum of the shrimps is serrated and has defensive function in some species. It can cause lacerations and bleeding. (Photo: Vidal Haddad Junior)

Fig. 2.75 The mantis shrimp (*Lysiosquilla* sp.) is an aggressive crustacean that can wound fishermen and aquarists with their sharp claws (see in the detail). (Photo: Vidal Haddad Junior)

Fig. 2.76 Wounds caused by mantis shrimp is not uncommon in fishermen, having potential to provoke serious lacerations. These animals are called "thumbspliters" in the Caribbean region. (Photo: João Luiz Costa Cardoso, dermatologist)

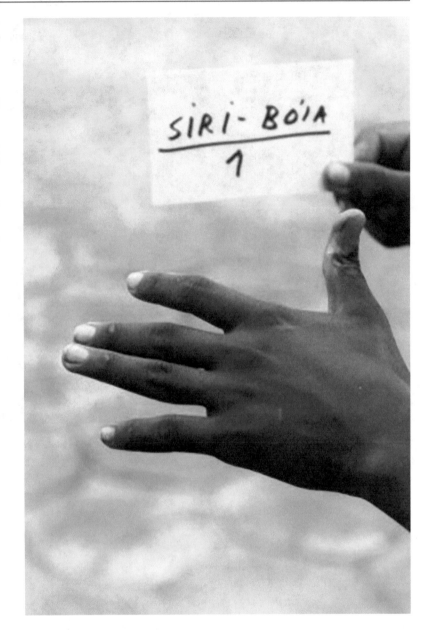

Fig. 2.77 The barnacles are fixed crustaceans that present sharp edges and causes incise wounds in bathers. (Photo: Vidal Haddad Junior)

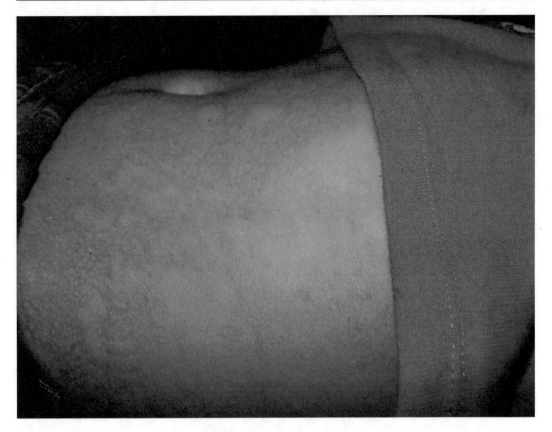

Fig. 2.78 Urticaria is one of the allergic manifestations caused by the ingestion of the meat of crustaceans. In this case, the victim the victim had eaten shrimp about 15 minutes before the onset of erythematous and edematous eruption. (Photo: Vidal Haddad Junior)

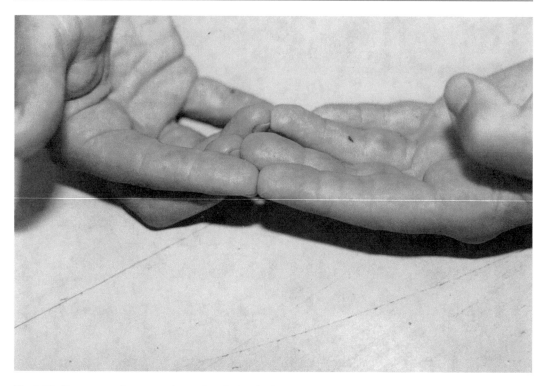

Fig. 2.79 Eczematous plaques are observed on the hands of shrimps and prawns cleaners and fishermen due an allergic and traumatic dermatitis. (Photo: Vidal Haddad Junior)

Phylum Insecta (Giant Water Bugs)

The Belostomatidae hemiptera are large venomous and carnivorous semi-aquatic arthropods (Fig. 2.80). They are found near or in freshwater environments, hunting other arthropods, small fish, and frogs. The giant water bugs can measure up to 10 cm in length, similar to the species *Lethocerus delpontei*, and those insects can cause envenomations in human beings and other animals, especially those human who work in small water streams. The bite is inflicted by a proboscis and provokes severe pain and moderate inflammation, due to the injection of enzymes and toxins of the digestive system of the animal (Fig. 2.81). There is a report of reversible paralysis of the limb of a victim. Some studies associate the sting of these insects with the emergence of Buruli ulcer, caused by *Mycobacterium ulcerans* [2, 50].

Fig. 2.80 Giant water bugs are venomous insects that live in freshwater pools, hunting fish and other small animals. These large arthropods can inject enzymes and toxins by a proboscis (see details) causing a local painful and inflammatory process with occasional paralysis of the member affected. (Photos: Vidal Haddad Junior)

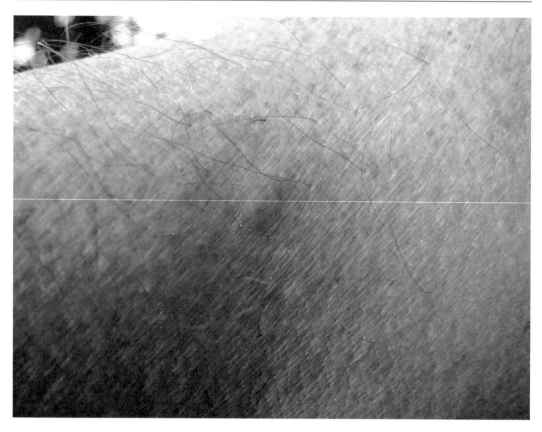

Fig. 2.81 The inflammation and erythematous papule were caused by a bite of a giant water bug in a researcher of fish in freshwater environments. (Photo: Vidal Haddad Junior)

References

1. Haddad V Jr. Avaliação Epidemiológica, Clínica e Terapêutica de Acidentes Provocados por Animais Peçonhentos Marinhos na Região Sudeste do Brasil (thesis). São Paulo (SP): Escola Paulista de Medicina; 1999. 144 pp.
2. Haddad V Jr. Atlas de animais aquáticos perigosos do Brasil: guia médico de diagnóstico e tratamento de acidentes (Atlas of dangerous aquatic animals of Brazil: a medical guide of diagnosis and treatment). São Paulo: Editora Roca; 2000. 148 pp.
3. Haddad V Jr. Animais aquáticos de importância médica. Rev Soc Bras Med Trop. 2003;36:591–7.
4. Haddad V Jr. Animais Aquáticos Potencialmente Perigosos do Brasil: Guia médico e biológico (Potentially Dangerous aquatic animals of Brazil: a medical and biological guide). São Paulo: Editora Roca; 2008. 268 pp.
5. Haddad V Jr, Lupi O, Lonza JP, Tyring SK. Tropical dermatology: marine and aquatic dermatology. J Am Acad Dermatol. 2009;61:733–50.
6. Burke WA. Coastal and marine dermatology. In: Forum in the Meeting of the American Academy of Dermatology. San Francisco, 1997.
7. Volkmer-Ribeiro C, Lenzi HL, Oréfice F, Pelajo-Machado M, de Alencar LM, Fonseca CF, Batista TCA, Manso PPA, Coelho J, Machado M. Freshwater sponge spicules: a new agent of ocular pathology. Mem Inst Oswaldo Cruz. 2006;101:899–903.
8. Halstead BW, Auerbach PS, Campbell DA. A colour atlas of dangerous marine animals. London: Wolfe Medical Publications; 1990. 192 pp.
9. Fisher AA. Atlas of aquatic dermatology. New York: Grume and Straton; 1978. 113 pp.
10. Meier J, White J. Clinical toxicology of animal venomous and poisonous. Florida: CRS Press; 1995. 504 pp.
11. Cazorla-Perfetti DJ, Lovo J, Lugo L, Acosta ME, Morales P, Haddad V Jr, Rodriguez-Moralles AJ. Epidemiology of the cnidarian *Physalia physalis* stings attended at a health care center in beaches of Adicora, Venezuela. Travel Med Infec Dis. 2012;10:263–6.
12. Haddad V Jr, Silveira FL, Cardoso JLC, Morandini AC. A report of 49 cases of cnidarian envenoming from southeastern Brazilian coastal waters. Toxicon. 2002;40:1445–50.
13. Haddad V Jr, Silva G, Rodrigues TC, Souza V. Injuries with high percentage of systemic find-

ings caused by the cubomedusa *Chiropsalmus quadrumanus* (Cnidaria) in Southeast region of Brazil: report of ten cases. Rev Soc Bras Med Trop. 2003;36:84–5.

14. Risk JY, Haddad V Jr, Cardoso JLC. Envenoming caused by a Portuguese man-o'-war (*Physalia physalis*) manifesting as purpuric papules. An Bras Dermatol. 2012;87:644–5.

15. Haddad V Jr, Migotto AE, Silveira FL. Skin lesions in envenoming by cnidarians (Portuguese man-of-war and jellyfish): etiology and severity of the accidents on the Brazilian Coast. Rev Inst Med Trop São Paulo. 2010;52:43–6.

16. Haddad V Jr, Virga R, Bechara A, Silveira FL, Morandini AC. An outbreak of Portuguese man-of-war (*Physalia physalis* - Linnaeus, 1758) envenoming in Southeastern Brazil. Rev Soc Bras Med Trop. 2013;46(5):641–4.

17. Burnett JW, Calton GJ, Burnett HW. Jellyfish envenomation syndromes. J Am Acad Dermatol. 1986;14:100–6.

18. Fenner PJ, Williamson JA. Worldwide deaths and severe envenomation from jellyfish stings. Med J Aust. 1996;165(11–12):658–61.

19. Lisa-Ann G. *Malo kingi*: a new species of Irukandji jellyfish (Cnidaria: Cubozoa: Carybdeida), possibly lethal to humans, form Queensland, Australia. Zootaxa. 2007;1659:55–68.

20. Tibballs J. Australian venomous jellyfish, envenomation syndromes, toxins and therapy. Toxicon. 2006;48(7):830–59.

21. Resgalla C Jr, Rossetto AL, Haddad V Jr. Report of an outbreak of stings caused by *Olindias sambaquiensis* MULLER, 1861 (Cnidaria:Hydrozoa) in Southern Brazil. Braz J Oceanogr. 2011;59:391–6.

22. Haddad V Jr, Cardoso JLC, Silveira FLS. Seabather's eruption: report of five cases in the Southeast Region of Brazil. Rev Inst Med Trop Sao Paulo. 2001;43:171–2.

23. Rossetto AL, Dellatorre G, Silveira FL, Haddad V Jr. Seabather's eruption: a clinical and epidemiological study of 38 cases in Santa Catarina State, Brazil. Rev Inst Med Trop Sao Paulo. 2009;51:169–75.

24. Rossetto AL, Mora JM, Correa PR, Resgalla C Jr, Proença LAO, Silveira FL, Haddad V Jr. Prurido do traje de banho: relato de seis casos no Sul do Brasil. Rev Soc Bras Med Trop. 2007;40:78–81.

25. Garcia PJ, Schein RMH, Burnett JW. Fulminant hepatic failure from a sea anemone sting. Ann Intern Med. 1994;120:665–6.

26. Marques AC, Haddad V Jr, Migotto AE. Envenomation by a benthic Hydrozoa (Cnidaria): the case of *Nemalecium lighti* (Haleciidae). Toxicon. 2002;40:213–5.

27. Burnett JW, Kumar S, Malecki JM, Szmant AM. The antibody response in seabather's eruption. Toxicon. 1995;33:95–104.

28. Loten C, Stokes B, Warsley D, Seymour JE, Jiang S, Isbistier GK. A randomized controlled trial of hot water (45°C) immersion versus ice packs for pain relief in bluebottle stings. Med J Aust. 2006;4:329–33.

29. Fenner P. Awareness, prevention and treatment of world-wide marine stings and bites. Conference in international life saving federation medical/rescue proceedings. Australia, 1997.

30. Guevara BEK, Dayri JF, Haddad V Jr. Delayed allergic dermatitis presenting as a keloid-like reaction caused by sting from an Indo-Pacific Portuguese man-o`-war (*Physalia utriculus*). Clin Exp Dermatol. 2017;42(2):182–4.

31. Haddad V Jr, Morandini AC, Rodrigues LE. Jellyfish blooms causing mass envenomations in aquatic marathonists: report of cases in S and SE Brazil (SW Atlantic Ocean). Wilderness Environ Med. 2018;29:142–5.

32. Pereira JCC, Spilzman D, Haddad V Jr. Anaphylactic reaction/angioedema associated with jellyfish sting. Rev Soc Bras Med Trop. 2018;51:115–7.

33. Haddad V Jr, Costa MAO, Nagata R. Outbreak of jellyfish envenomations caused by the species *Olindias sambaquiensis* (Cnidaria: Hydrozoa) in the Rio Grande do Sul State (Brazil). Rev Soc Bras Med Trop. 2019;52:e20190137.

34. Haddad V Jr, Morandini AC. Jellyfish stings. In: World atlas of jellyfish. Berlin: Dölling und Galitz Verlag; 2019. 816 pp.

35. Rossetto AL, Cruz CCB, Pereira ICC, Nunes JA, Martins MM, Nicolacópulos T, Rosetto AL, Haddad V Jr. Diagnostic confusion between seabather's eruption as well as dermatophytosis and parasitic infestations. Rev Soc Bras Med Trop. 2020;53:e20190462.

36. Rossetto AL, Venzon SL, Cruz CCB, Dimatos OC, Rossetto AL, Morandini AC, Haddad Jr V. Seabather's eruption: description of a new clinical manifestation. J Eur Acad Dermatol Venereol. 2020. https://doi.org/10.1111/jdv.16537. PMID: 32342539.

37. Rossetto AL, Rossetto AL, Guevara BEK, Haddad V Jr. Seabather's eruption presents Koebner phenomenon? J Eur Acad Dermatol Venereol. 2019;34(2):e93–5.

38. Guevara BEK, Dayrit JF, Haddad V Jr. Seabather's eruption caused by the thimble jellyfish (*Linuche aquila*) in the Philippines. Clin Exp Med. 2017;42(7):808–10.

39. Bastos DMRF, Ferreira DMR, Haddad V Jr, Nunes JLS. Human envenomations caused by Portuguese man-of-war (*Physalia physalis*) in urban beaches of São Luis City, Maranhão State, Northeast Coast of Brazil. Rev Soc Bras Med Trop. 2017;50:130–4.

40. Aquino GGES, Haddad V Jr, Pires VA. Avaliação dos acidentes ocorridos por cnidários no município de Salinópolis/Pará (Brasil). Biota Amazônia. 2019;49:37–40.

41. Haddad V Jr, Novaes SPMS, Miot HA, Zuccon A. Accidents caused by sea urchins – the efficacy of precocious removal of the spines in the prevention of complications. An Bras Dermatol. 2002;77:123–8.

42. Rossetto AL, Mota JM, Haddad V Jr. Sea urchin granuloma. Rev Inst Med Trop Sao Paulo. 2006;48:303–6.

43. Haddad V Jr. Observation of initial clinical mani-
    festations and repercussions from the treatment of
    314 human injuries caused by black sea urchins
    (*Echinometra lucunter*) on the southeastern Brazilian
    coast. Rev Soc Bras Med Trop. 2012;45:390–2.
44. Haddad V Jr, Paula Neto JB, Cobo VJ. Venomous
    mollusks: the risks of human accidents by *Conus*
    snails (Gastropoda, Conidae) in Brazil. Rev Soc Bras
    Med Trop. 2006;39(5):498–500.
45. Haddad V Jr, Coltro M, Simone LRL. Report of a
    human accident caused by *Conus regius* (Gastropoda,
    Conidae). Rev Soc Bras Med Trop. 2009;42:446–8.
46. Haddad V Jr, Moura R. Acute neuromuscular mani-
    festations in a patient associating with ingesting
    octopus (*Octopus* sp). Rev Inst Med Trop Sao Paulo.
    2007;49:59–61.
47. Haddad V Jr, Magalhães CA. Infiltrated plaques
    resulting from an injury caused by the common octo-
    pus (*Octopus vulgaris*): a case report. J Venom Anim
    Toxins Incl Trop Dis. 2014;20:1–2.
48. Haddad V Jr, Freire FAM, Joustra JPL. Suction
    purpura in humans caused by octopus arms. Int J
    Dermatol. 2014;53(3):e174–5.
49. Amaral ALS. Mantis shrimp, Siriboia or tambu-
    rutaca (Crustacea: Stomatopoda): morphology of
    raptorial claws and its relation with human injuries
    (thesis). Botucatu: Pós-Graduação em Zoologia,
    Instituto de Biociências, Universidade Estadual
    Paulista; 2020.
50. Haddad V Jr, Schwartz ENF, Schwartz CA, Carvalho
    LN. Bites caused by giant water bugs belonging to
    Belostomatidae family (Hemiptera, Heteroptera) in
    humans: a report of seven cases. Wilderness Environ
    Med. 2010;21:130–3.

Injuries by Aquatic Vertebrate Animals

3

The Phylum Chordata (Fish and Reptiles)

Wounds and envenomations caused by fish are events that can occur in certain populations, sometimes being characterized as occupational accidents. A typical example involves professional fishermen and their families, who may be injured while cleaning and preparing fish. Such injuries commonly occur on the hands (mainly) and on the feet of the victims. Individuals who engage in recreational or sport fishing are also exposed to these risks. Another affected group is bathers, and the most common circumstances of these injuries involve small fish (especially marine catfish) being discarded on sandy beaches or in shallow water by amateur or professional fishermen because these fish are of little interest for consumption or trade. However, the fish's venom can remain active for about 24 hours after its death [1–9].

The Class Chondrichthyes (Cartilaginous Fish)

Marine Stingrays

Stingrays are elasmobranch fish with a cartilaginous skeleton and adapted fins that resemble wings and facilitate underwater locomotion. Not all stingrays found in marine environments have stingers, but stingrays are capable of causing trauma and envenomation. The stingrays most commonly associated with human injuries belong to the Dasyatidae, Myliobatidae, Rhinopteridae, and Gymnuridae families (Figs. 3.1, 3.2, 3.3, 3.4, 3.5, 3.6, 3.7, 3.8, 3.9, and 3.10) [10].

The marine stingrays most commonly associated with injuries in humans can carry 1–4 venomous stings on their tail (Figs. 3.11 and 3.12). The stinger is composed of dentin and, depending on the size of the ray, it can reach a length of up to 20 cm. The morphology of serrated edges can cause major lacerations, especially during extraction of the stinger from the wound. The stinger breaks the sheath that covers it when it penetrates the victim, and the contents of the glandular tissue that occupies the groove in the ventral position of the stinger flow into the wound, causing localized envenomation. The venom contains high molecular weight polypeptides—such as serotonin, phosphodiesterase, 5-nucleotidase, and hyaluronidases—which are responsible for neurotoxic activity (severe pain), cardiotoxic activity, and possibly skin necrosis at the site of the sting [1–9].

The glands producing the venom are not individualized but dispersed over the stinger, located in the tail. Injuries caused by stingrays always provoke intense pain, which can be excruciating, at the site of the sting. The pain can cause behavioral changes and somatic disorders (such as extreme anxiety), a rapid heartbeat, cold sweats, nausea/vomiting, and sphincter release. The

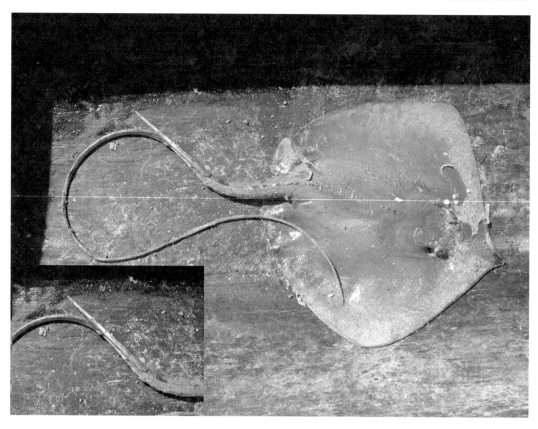

Fig. 3.1 A stingray of the Dasyatidae family on the sand (*Dasyatis guttata*). The whiptail stingray is probably the most common ray causing human envenomations. (Photo: Vidal Haddad Jr.)

wound site becomes edematous, and pallor can occur in the area surrounding the perforation. That area becomes erythematous and warm, evidencing the inflammation caused by the venom (Figs. 3.13 and 3.14a, b). Skin necrosis can be observed but is not common (Figs. 3.15 and 3.16). Occasionally, a bacterial infection sets in, with serious consequences. Some fatalities have been described, caused by large animals and associated with thoracic or abdominal cavity perforations by a stinger [1–10].

The immediate treatment for pain control, which is accessible even to the victims themselves, is immersion of the wound area (which is usually located on an extremity) in hot water. The water should not be scalding, and it must be kept in mind that because the patient has changes in their thermal sensitivity at the site of the sting, it is important that the temperature be tested by another person to prevent burns.

All fish venom is thermolabile, and application of hot water interferes with the activity of the venom and tends to inactivate it, but the anesthetic effect seems more linked to the vasodilatation that hot water triggers. Studies of stingray venom and other fish venoms have shown that they cause intense vasoconstriction, and this effect, in turn, causes tissue ischemia, pallor, cyanosis, and, more rarely, skin necrosis [10]. It is possible to achieve significant pain relief with local immersion in hot water, but the pain returns when the patient removes the affected area from the water [4, 5, 10].

Other measures, such as withdrawal of stinger fragments from the wound and intensive cleaning of lacerated wounds, must be done in a hospital by a health professional. Surgery may be required for extraction of fragments (Fig. 3.17).

Some rays (those of the Torpedinidae family and of the *Narcine* and *Torpedo* genera) can deliver electric shocks ranging from 15 to 200 V,

Fig. 3.2 Stinger of the longnose stingray showed in Fig. 3.1 in detail. The stinger of Dasyatidae family is very dangerous, for the envenomation and for the trauma pre- disposed by the position of the stinger and the length of the tail. (Photo: Vidal Haddad Jr.)

but these accidents are very rare and have only minor consequences for humans (Fig. 3.18) [4, 5].

The sawfish (*Pristis pectinata*) has a saw-shaped "beak," which can cause serious lacerated wounds in humans. This is a real risk, but occur- rences have been rare and poorly documented. Injuries by sawfish should be treated similarly to other severe trauma, with special attention to copious bleeding and internal organ injuries (Figs. 3.19 and 3.20) [4, 5].

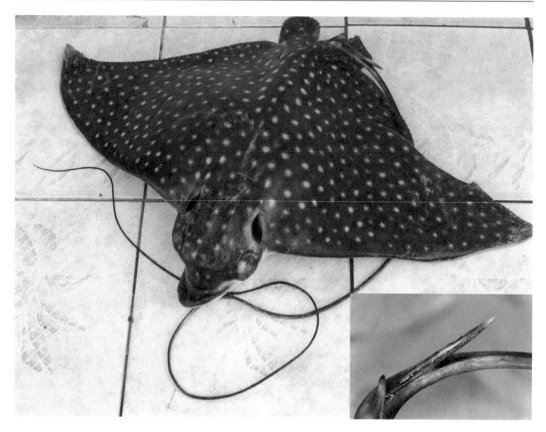

Fig. 3.3 The spotted stingrays (*Aetobatus narinari*) are stingrays of the Myliobatidae family (eagle rays). They are common in warm waters around the world and can cause severe envenomation in human beings. (Photo: Vidal Haddad Jr.)

Fig. 3.4 Stinger of the *Aetobatus narinari* stingray. (Photo: Vidal Haddad Jr.)

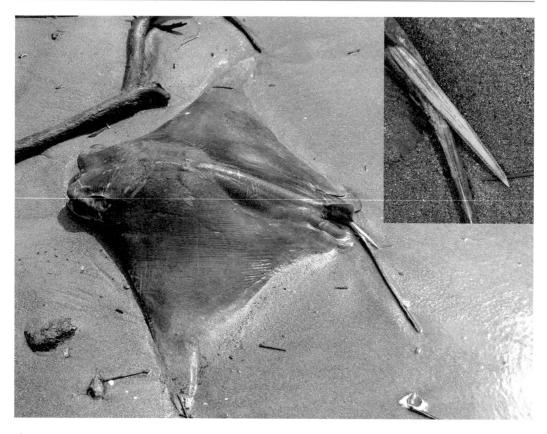

Fig. 3.5 The Rhinopteridae family of stingrays (the cownose stingrays) also can provoke severe injuries in humans. The species *Rhinoptera bonasus* is showed in the image. (Photo: Vidal Haddad Jr.)

Fig. 3.6 The stingrays of the Rhinopteridae family, as the other tem families of stingrays, can present one to four stingers in the tail. (Photo: Vidal Haddad Jr.)

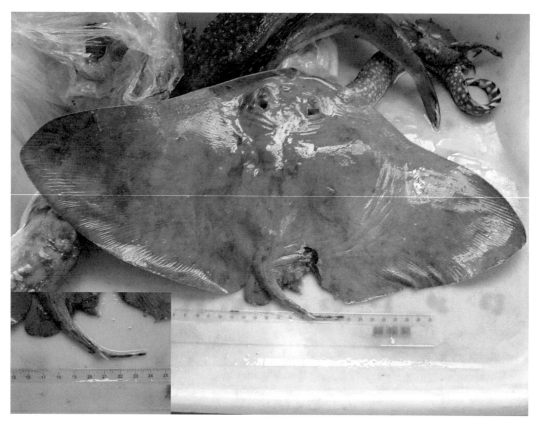

Fig. 3.7 The Gymnuridae family (butterfly stingrays) presents common stingrays with short tail. (Photo: Vidal Haddad Jr.)

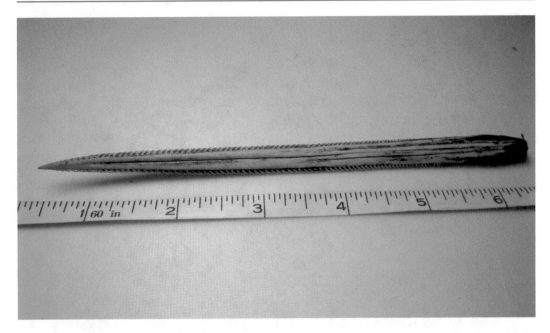

Fig. 3.8 Stingrays have stingers capable to provoke severe envenomations

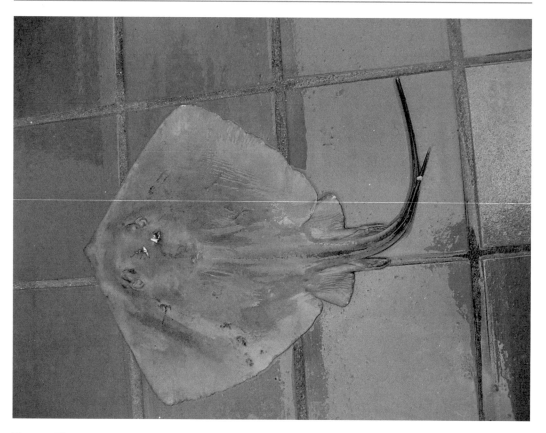

Fig. 3.9 The common stingray (*Dasyatis pastinaca*) is the most common stingray in the Eastern Atlantic and Mediterranean Sea, causing injuries in Europe and Africa. (Photo: Prof. João Pedro Barreiros Azores University, Portugal)

Fig. 3.10 The species *Myliobatis aquila* is the Common Eagle ray. They live also in Eastern Atlantic Ocean and Mediterranean Sea. (Photo: Prof. João Pedro Barreiros Azores University, Portugal)

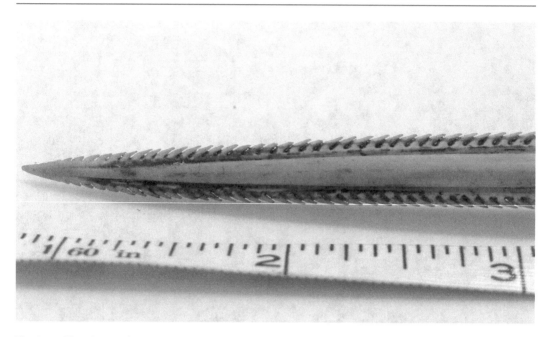

Fig. 3.11 The stingers of marine stingrays is very specialized tool for provoke trauma and inject the venom. (Photo: Vidal Haddad Jr.)

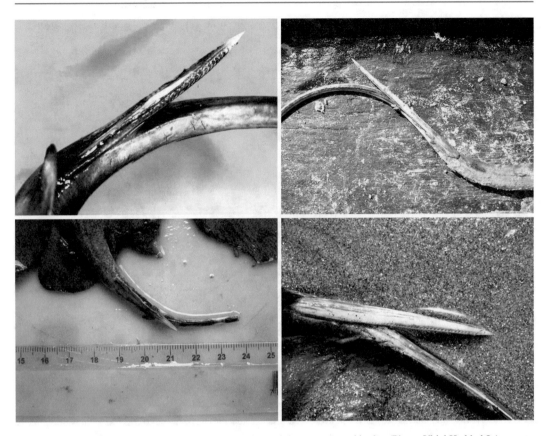

Fig. 3.12 Detail of the stingers showing the sharp point and the retroserrated barbs. (Photo: Vidal Haddad Jr.)

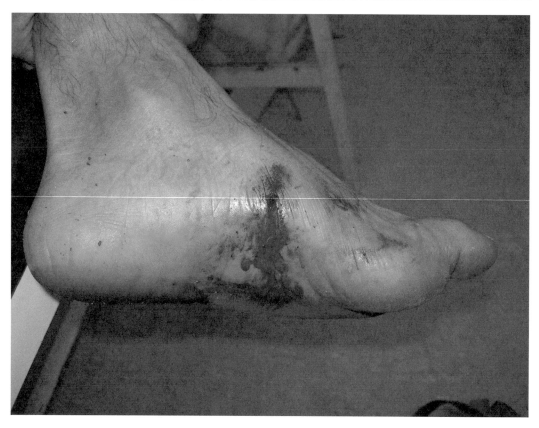

Fig. 3.13 The perforation with bleeding of this patient was registered minutes after the sting. The patient presented intense pain and local inflammation. (Photo: Vidal Haddad Jr.)

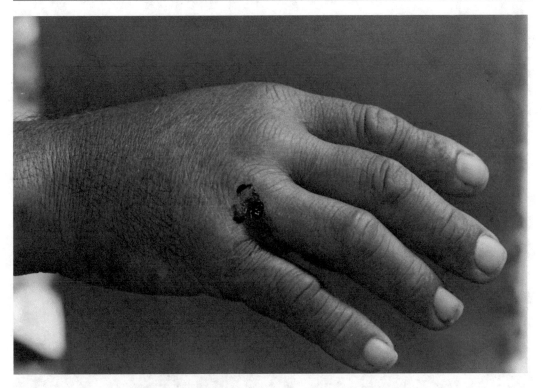

Fig. 3.14 (**a**) Intense inflammation observed in the hand of a fisherman after the sting of a Rhinopteridae stingray. (Photo: Vidal Haddad Jr.). (**b**) The perforation, caused by a Dasyatidae stingray shows alterations of colors on the foot of a bather. The pain was intense. (Photo: Vidal Haddad Jr.)

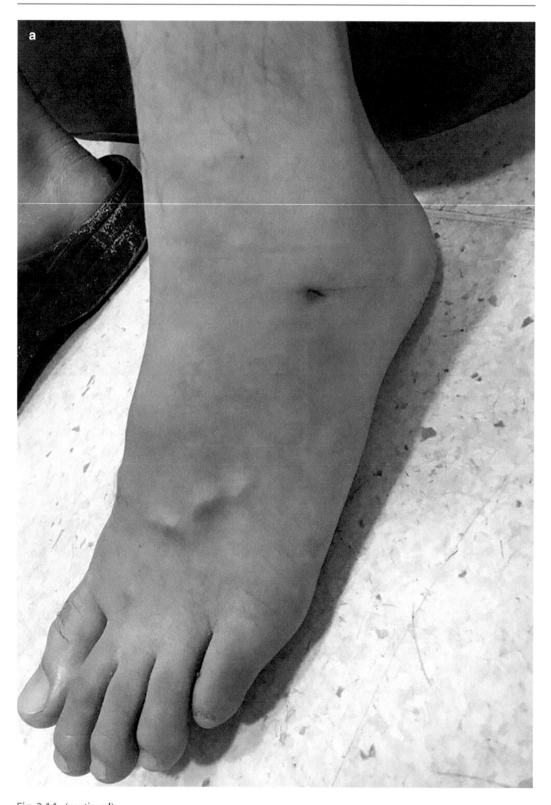

Fig. 3.14 (continued)

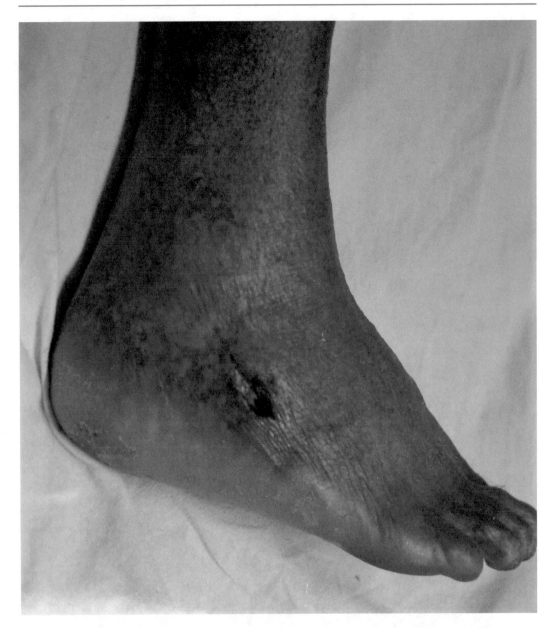

Fig. 3.15 Chronic ulcer in the left foot of a fisherman that stepped in a marine stingray on the boat. (Photo: Vidal Haddad Jr.)

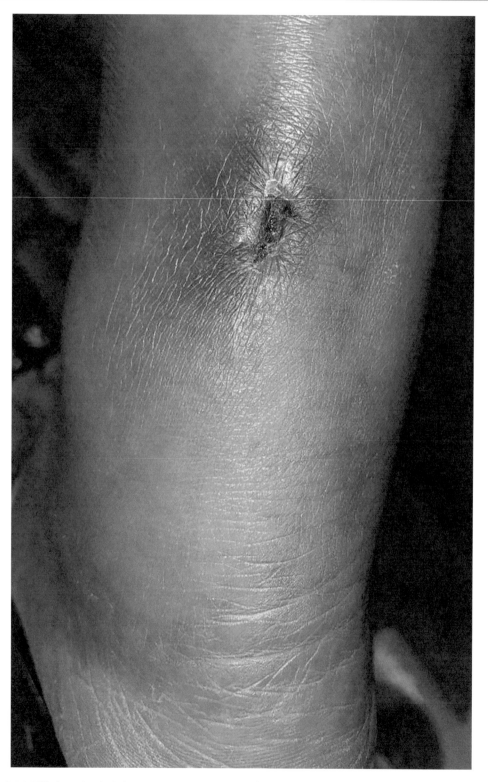

Fig. 3.16 This is a classic lesion caused by a marine stingray. The perforation usually occurs in the foot or ankle when the animal is stepped on and reacts. The edema and erythema are visible, but there is no marked necrosis. (Photo: Vidal Haddad Jr.)

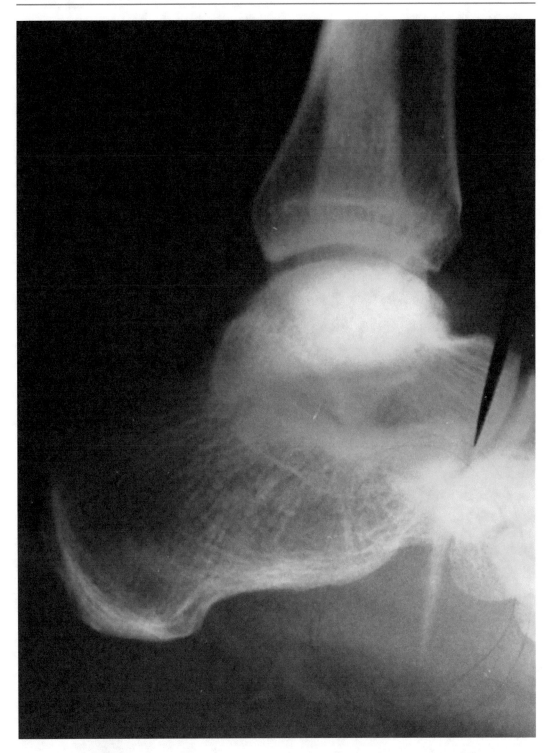

Fig. 3.17 Fragments of stingers of stingrays and catfish can break in the wound and provoke chronic inflammation and foreign body granulomas. Surgery is necessary to correct this complication. (Photo: Vidal Haddad Jr.)

Fig. 3.18 The marine electric rays are classified in the Torpedinidae family, *Narcine* and *Torpedo* genera. The shocks do not cause deaths (about 50 V). (Photo: Vidal Haddad Jr.)

Fig. 3.19 Sawfish are part of the Pristidae family. The species *Pristis pectinata* an cause severe laceration with their serrated "beak". (Photo: Prof. Ivan Sazima, zoologist)

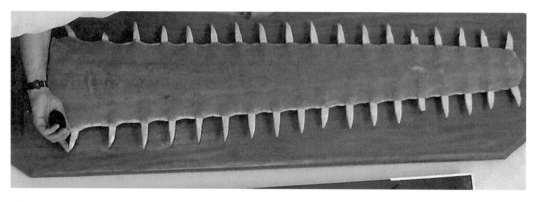

Fig. 3.20 The "saw" of the *Pristis pectinata* is a powerful weapon for defense and attack of the fish. (Photo: Vidal Haddad Jr.)

Freshwater Stingrays

Some species of marine stingray are adapted to freshwater environments. There are greater or lesser degrees of adaptation, ranging from true marine species that penetrate into rivers over long distances to other species (such as rays of the Potamotrygonidae family) that are fully adapted to live in freshwater and do not return to the sea. These stingrays are widely distributed throughout the river basins of South America, with appearances and habits similar to those of marine stingrays. This family comprises four genera (*Potamotrygon*, *Paratrygon*, *Plesiotrygon*, and *Heliotrygon*) and several species. *Potamotrygon motoro* is the species most widely distributed across the continent and, probably because of this, it causes the most injuries (Figs. 3.21, 3.22 and 3.23a–d). These stingrays remain buried in sand or mud in shallow water and cause many envenomations in people crossing rivers during the dry season [10–14].

On their tail, these rays may have 1–4 stingers, which penetrate the victim when the tail performs a whip movement and fires them at the victim. This is done to protect the fish, usually because it has been stepped on (Figs. 3.24, 3.25 and 3.26). *Stingrays do not attack; they only defend themselves.*

The immediate effects of the sting are localized and similar to those observed in marine stingrays, causing intense pain, edema, and erythema at the site of envenomation (Figs. 3.27 and 3.28). After approximately 24 hours, the pain subsides. As a rule, the sting tends to cause localized tissue necrosis (as has occurred in about 400 injuries observed by the author) (Figs. 3.29 and 3.30). The necrosis causes ulcers that are difficult to heal (Fig. 3.31a–c). The stings of freshwater stingrays provoke skin necrosis more frequently than those of marine stingrays. This fact seems to be associated with the action of proteolytic enzymes (especially hyaluronidases), which are present in high concentrations in the venom of river stingrays [10–15]. Secondary infection can precipitate or aggravate the ulcers, which are usually on the lower limbs, especially on an ankle or a foot (Fig. 3.36) [10]. Other complications are retention of fragments of the stinger, which can cause chronic inflammation at the site of the sting (Fig. 3.37), and extensive scarring resulting from the ulcers, which can cause skin cancer and restriction of bodily movement (Fig. 3.38).

Treatment involves the same initial steps for envenomation by any species of stingray (or other fish); the most important measure is immersion of the wound site in hot water, which causes vasodilatation and decreases pain. Use of systemic antibiotics should be encouraged for all lacerated wounds to prevent or control bacterial infections, which interfere with the evolution of the envenomation. The wound should be washed thoroughly and explored to remove fragments. Ulcers that are already established may require skin

Fig. 3.21 The freshwater stingrays of the Potamotrygonidae family are beautiful fish that live in South American rivers. (Photo: Prof. Domingos Garrone Neto, Registro Campus, São Paulo State University

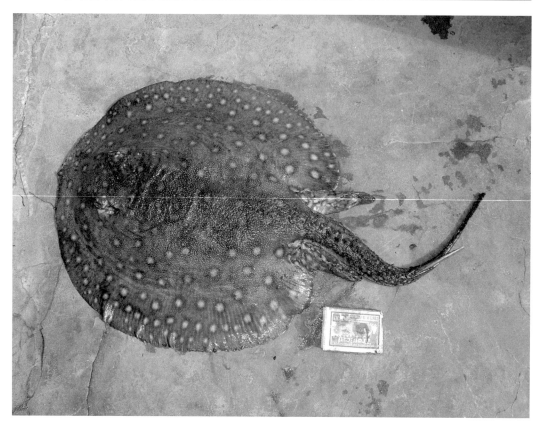

Fig. 3.22 *Potamotrygon motoro* is the most common species of freshwater stingrays in South America, causing the majority of the human injuries in the riverside population and fishermen. (Photo: Vidal Haddad Jr.)

grafts. Early excision of the lesion may be attempted empirically, but the results of this approach are dubious [10].

Injuries caused by stingrays are very common, as can be verified through interviews and observations of marine fishermen in coastal areas and freshwater fishermen throughout South America. Freshwater stingrays are now present in vast areas of Brazil where they were not previously observed [10].

Fig. 3.23 (**a**) Among various other species, *Potamotrygon falkneri* lives in Midwest Brazil and Argentina. (Photo: Vidal Haddad Jr.). (**b**) *Potamotrygon* aff. *P. wallacei*, the arraia-cururu, commom in the Rio Negro, Amazon region. (Photo: Vidal Haddad Jr.). (**c**) *Potamotrygon leopoldi*, the beautiful freshwater stingray of the Xingu Basin. Note the stinger in the detail. (Photo: Vidal Haddad Jr.). (**d**) Freshwater stingrays remain semi-buried in sandy bottoms waiting for preys. When stepped on, they whip the tail and intrude the stingers 1–4 on the victim. (Photo: Vidal Haddad Jr.)

Fig. 3.23 (continued)

Fig. 3.23 (continued)

Fig. 3.24 Stinger of *Potamotrygon falkneri* with the sheath covering the bony sting. The venomous tissue is in grooves and in the dark mucous on the sting. "Charged" stingers present this mucous. (Photo: Vidal Haddad Jr.)

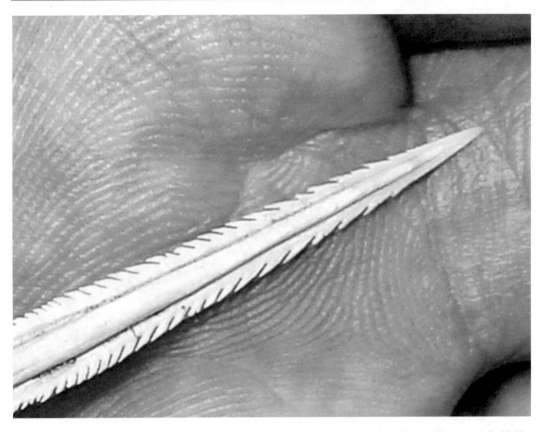

Fig. 3.25 "Discharged" stingers of *Potamotrygon falkneri*. Stings without venom in the barb still can provoke highly traumatic lesions. (Photo: Vidal Haddad Jr.)

Fig. 3.26 The "discharged" stingers cause deep wounds, but no envenomation

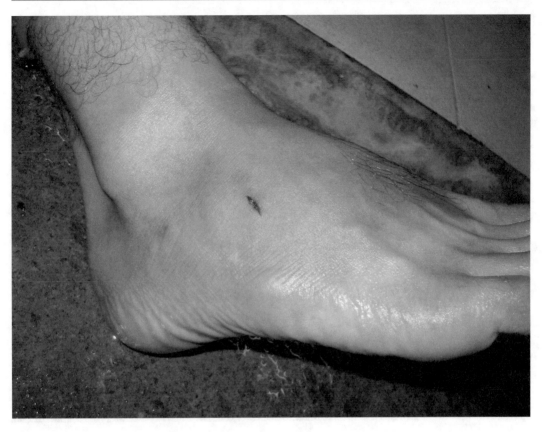

Fig. 3.27 A recent sting in a fisherman (two hours). The patient presented intense pain, but the inflammatory process still is discrete. (Photo: Vidal Haddad Jr.)

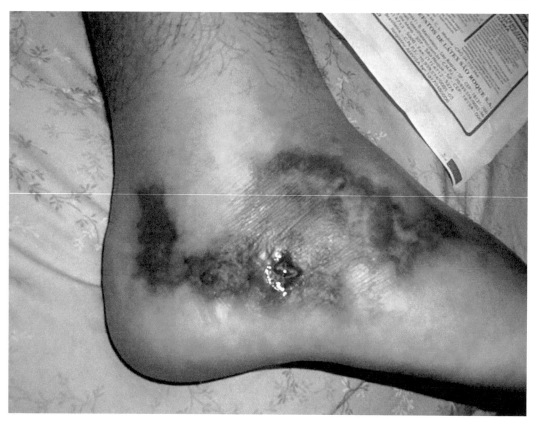

Fig. 3.28 This patient was stinged by a freshwater stingray three days after the observation and developed intense pain, marked inflammation and skin necrosis with a "map" aspect. (Photo: Vidal Haddad Jr.)

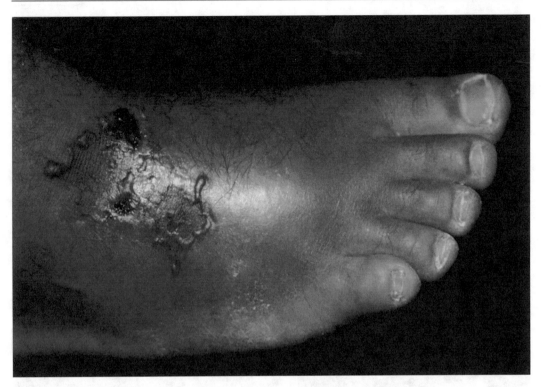

Fig. 3.29 The skin necrosis "in map" is characteristic of the freshwater stingray's envenomation and occurs due the movement of the venom by gravity form the point of the perforation. (Photo: Vidal Haddad Jr.)

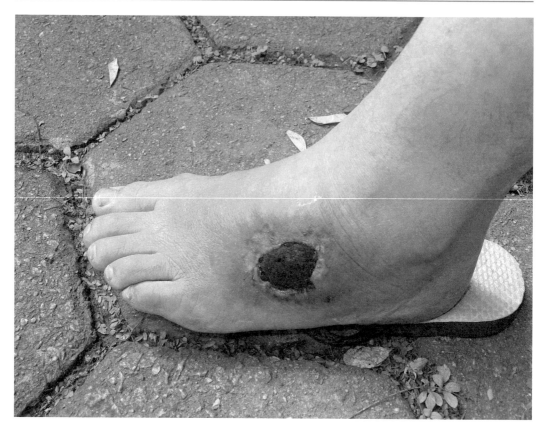

Fig. 3.30 Dry skin necrosis caused by a sting of *Potamotrygon motoro* in Southeast region of Brazil. The scar is firmly adhered to skin. (Photo: Vidal Haddad Jr.)

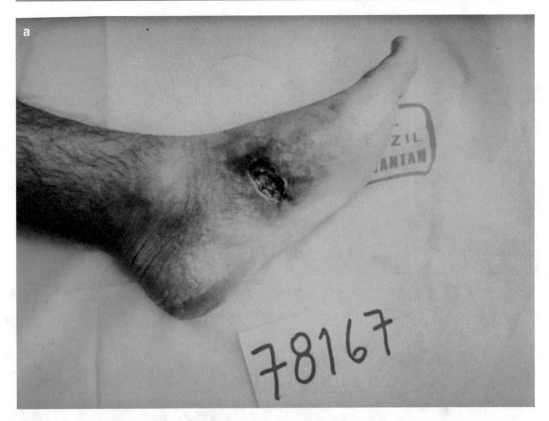

Fig. 3.31 (**a**) The envenomation by freshwater stingrays always causes skin necrosis and deep ulcers, which can persist for months, decreasing the work capacity of the victims. (Photo: Vidal Haddad Jr.). (**b**) Stings from freshwater stingrays also usually happen on the ankle or foot of the victims. Deep perforation and immediate intense pain are characteristic of the envenomation. Necrosis sets in later and is marked by an anterior ischemic area. (Photos: Vidal Haddad Jr.). (**c**) On tourist fishing routes in the Brazilian Pantanal, live baits are sold to tourists. As these crustaceans and small fish are caught with sieves, there is exposure to rays, which often cause accidents to workers. (Photo: Vidal Haddad Jr.)

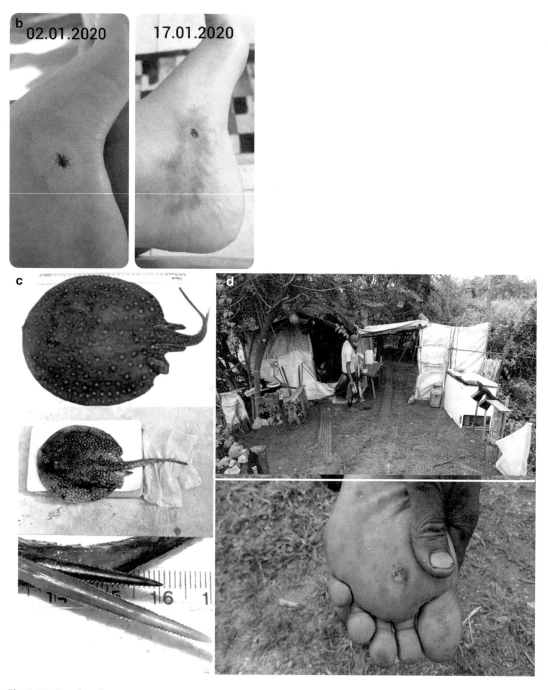

Fig. 3.31 (continued)

Box 3.1 Stingrays

Case 1: A 62-year-old fisherman reported having suffered a sting by a "yellow stingray" on the dorsum of his left hand while removing shrimp from a fishing net on a boat at sea (Fig. 3.32). When intense pain set in, he etched the spot with a hot knife, but the pain only decreased after four hours, and his symptoms of localized edema and erythema persisted. He was treated with medicaments that did not know refer. He said that he then avoided alcohol and contact with women for 6 months to avoid wound complications. On examination, he had a deep ulcer with a purulent secretion, which was caused by skin necrosis at the site of the sting and took 4 months to heal (Fig. 3.33).

Case 2: A 43-year-old man was injured in the Pardo River (a tributary of the Paraná River, Plata Basin, Brazil). He was stung on his left leg while removing fish from fish hooks left in the river (Fig. 3.34) and collapsed on the ground immediately because of the pain (Fig. 3.35). He also experienced cold sweats and malaise. The next day, the site of the sting was erythematous and pale. After 15 days, the site formed a "black skin" with a strong smell. After 1 month, he had a 10 cm ulcer, which took months to heal and was treated with traditional medicine prepared by a pharmacist.

Comments: These two cases are typical of stingray envenomations, which cause severe pain. Skin necrosis occurs invariably after freshwater stingray stings but only occasionally after marine stingray stings. Although the venom causes only localized effects, the pain and skin necrosis are severe, and there are only nonspecific treatment options in the early and late stages after envenomation.

Fig. 3.32 Sandpaper stingray, the yellow stingray of the Dasyatis (Hypanus) genus. (Photo: Vidal Haddad Jr.)

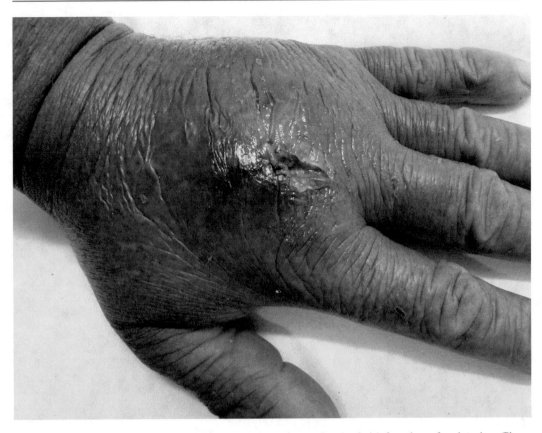

Fig. 3.33 The secondary infection is a complication in envenomations caused by freshwater stingrays. This image shows a severe infection caused by *Aeromonas hydrophila* after a sting (exclude) four days after the sting. (Photo: Vidal Haddad Jr.)

Fig. 3.34 The ocellated freshwater stingray (*Potamotrygon motoro*) is the most comem species of freshwater stingray in the South America. (Photo: Vidal Haddad Jr.)

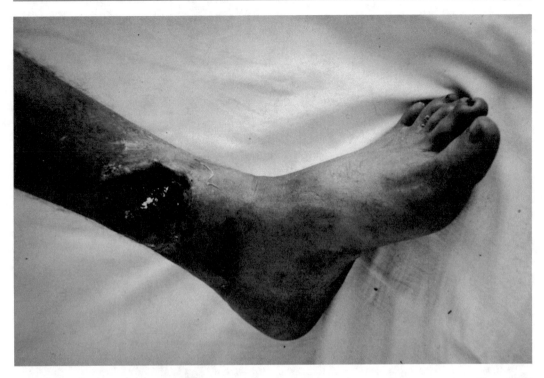

Fig. 3.35 A great ulcer developed after the initial necrosis caused by the freshwater stingray´s venom. (Photo: Vidal Haddad Jr.)

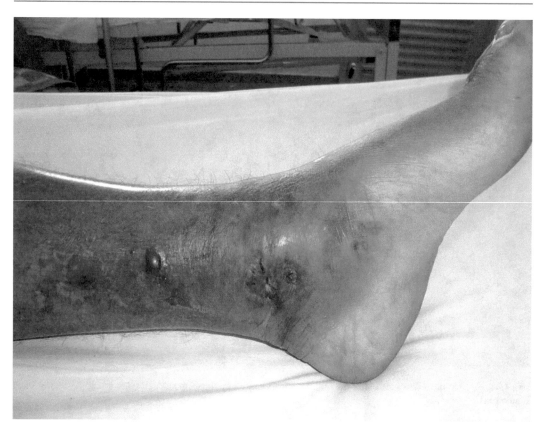

Fig. 3.36 The secondary infection is a complication in envenomations caused by freshwater stingrays. This image shows a severe infection caused by *Aeromonas hydrophila* after a sting four days after the sting. (Photo: Vidal Haddad Jr.)

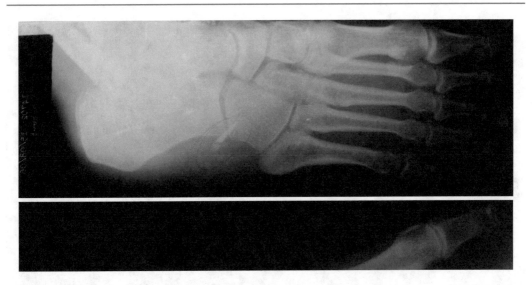

Fig. 3.37 The simple radiologic exam shows a fragment of a stinger of stingray in the foot of a victim. This is a common complication and the patients should be evaluated for the possibility. (Photo: Vidal Haddad Jr.)

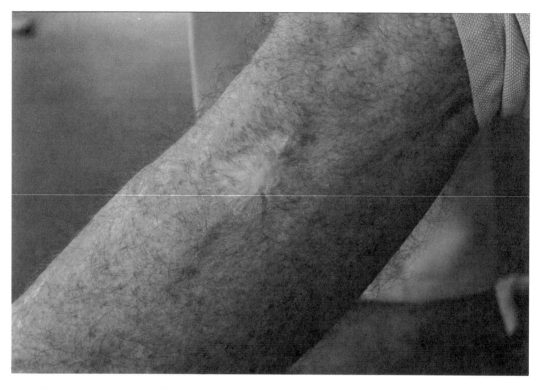

Fig. 3.38 A large scar is the result of skin necrosis and ulcers, and may cause persistent physical limitations in the victim. (Photo: Vidal Haddad Jr.)

Sharks

(This section was written in collaboration with Prof. Otto Bismarck Fazzano Gadig from the Instituto de Biociências, Campus do Litoral Paulista, São Paulo State University [UNESP], São Paulo, São Paulo State, Brazil.)

Although sharks have been implicated in human deaths and exploited exhaustively because of media fears, attacks (which is a questionable term from the behavioral point of view) by large sharks are unusual episodes, even in some areas such as Australia, the Pacific coast of the USA, South Africa, and the metropolitan area of Recife in northeastern Brazil, where they have been highly publicized [16–18] (Fig. 3.39).

Injuries inflicted by large shark bites are usually serious and can cause death, especially those caused by the species *Carcharhinus leucas* (the bull shark), *Galeocerdo cuvier* (the tiger shark) and *Carcharodon carcharias* (the great white

shark) (Figs. 3.40, 3.41, 3.42, and 3.43). The bull shark is probably the main species responsible for human attacks, as this species is widely distributed in tropical and subtropical inshore waters, and it can travel thousands of kilometers in large river systems and lakes, including the Amazon River, Nicaragua Lake, Mississippi River, and Ganges River. However, the tiger shark is also associated with attacks in coastal and insular tropical areas, and the great white shark is known for injuring victims in temperate regions, such as California, South Africa, and Australia [2, 4, 7].

Clear shark bites usually have a half-moon shape and marks in parallel rows caused during perforation by the teeth, but the pattern depends on the bite kinematics, the species involved, and several other aspects related to the attack. The bite may be very traumatic, causing large lacerations and extensive destruction and loss of body tissue [17] (Figs. 3.44 and 3.45). With regard to

Fig. 3.39 An alert regarding the risk of attacks by sharks in the metropolitan area of Recife (Pernambuco State, Brazil). This area has one of the highest rates of attacks in the world. (Photo: Vidal Haddad Jr.)

the two most dangerous large tropical coastal sharks (bull and tiger sharks), the lacerations usually result from multiple bites, and the major wounds are caused by the upper triangular and serrated teeth, resulting in an irregular and undefined pattern. Injuries caused by great white sharks can, in many cases, be proportionally less severe, since this species exhibits investigative preattack behavior that results in a single bite. Human deaths caused by shark attack events are related to large vessel injuries (which are most often caused by the first bite and/or the second bite), bleeding, and consequent hypovolemic shock.

Given such variations in wound characteristics, the clinical approach in shark attack treatment depends on the lesions. The major aim is to stop arterial bleeding, which must be interrupted or reduced by direct pressure on the vessel with a clean cloth—a fundamental measure in the case

of large lesions—or even by indirect pressure to avoid blood losses from several lesions.

Extensive injuries cause severe bleeding and hypovolemic shock, which are major problems and must be treated as a priority, with special attention to stabilization of breathing and circulation. The initial measures include oxygen administration and obtaining a route for administration of intravenous fluids and blood transfusion. After these first steps, it is necessary to start specialized treatment and hospital care [17, 18].

The injuries can be sutured, but the closed edges will always predispose the patient toward bacterial infections. The devitalized tissue should be debrided gently, and the area should be washed thoroughly to reduce the risk of bacterial infection.

Bacterial infection (usually by anaerobic bacteria) is very common, and preventive antibiotics should be used as a rule. In the author's experience, extensive wounds and all lacerations

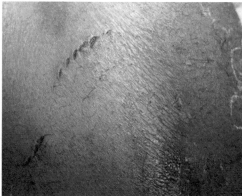

Fig. 3.40 *Top:* Worldwide, the bull shark (*Carcharhinus leucas*) is probably the most frequent cause of shark attacks on humans. *Bottom:* The great white shark (*Carcharodon carcharias*), and detail of a shark's bite on a bather. (Photos: Prof. Otto Bismarck Fazzano Gadig, Instituto de Biociências, Campus do Litoral Paulista, São Paulo State University [UNESP], São Paulo, São Paulo State, Brazil)

incurred in an aquatic environment are highly susceptible to infection, and antibiotics may be helpful in preventing infections, which can be severe (see section "Treatment of Trauma and Envenomations by Fish"). A careful search must be conducted for fragments of teeth or other materials in the wound. These measures are very important for treatment of both minor and extensive injuries [2, 4, 7, 18].

The spiny dogfish shark genus *Squalus* has dorsal fin spines (as do numerous other species). These spines are covered by a venomous glandular epithelium, the composition of which is not well known (Figs. 3.46 and 3.47) [4]. Pain is the main symptom of injury caused by these spines. The authors have previously had an opportunity to observe two cases of stings in professional fishermen, and the main symptom was moderate pain. The symptoms ceased about 1 hour after the injury [19]. Like *Squalus*, Indo-Pacific sharks of the genus *Heterodontus* have venomous dorsal spines. The elephant fish, or chimera (*Callorhinchus callorhynchus* and others), features a large venomous dorsal spike, but human injuries from this are very rare (Figs. 3.48 and 3.49).

Fig. 3.41 The bull shark can travel over large distances in freshwater environments. (Photo: Prof. Otto Bismarck Fazzano Gadig, Instituto de Biociências, Campus do Litoral Paulista, São Paulo State University [UNESP], São Paulo, São Paulo State, Brazil)

Fig. 3.42 Tiger sharks are potentially aggressive against humans. Occasionally, these big fish come close to fishing boats and are eventually captured. (Photo: Vidal Haddad Jr.)

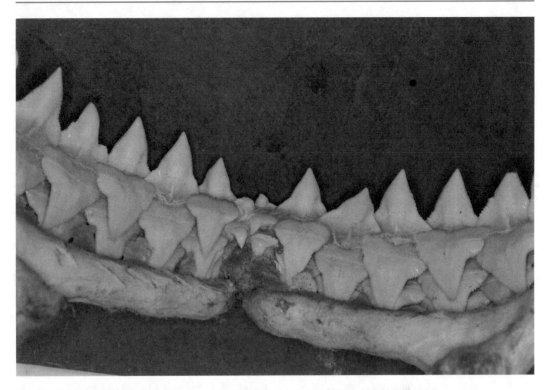

Fig. 3.43 Dental arch of a shark, showing rows of replacement teeth. (Photo: Vidal Haddad Jr.)

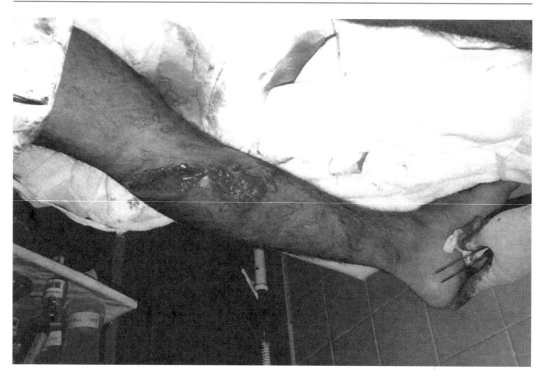

Fig. 3.44 Extensive lacerations and severe bleeding are observed in a bather after a great white shark attack in Brazil. (Photo: Prof. Otto Bismarck Fazzano Gadig, Instituto de Biociências, Campus do Litoral Paulista, São Paulo State University [UNESP], São Paulo, São Paulo State, Brazil)

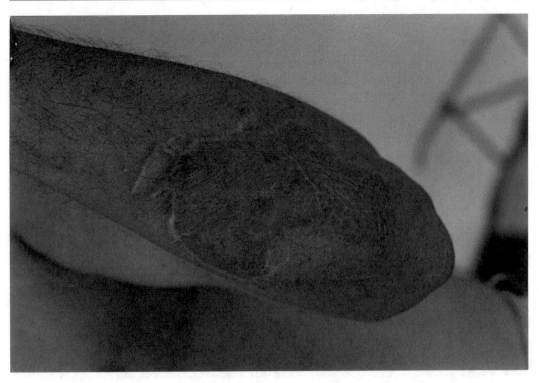

Fig. 3.45 These extensive scars are the result of a bite from a bull shark, which attacked a fisherman bringing in his fishing nets near a beach. (Photo: Vidal Haddad Jr.)

Fig. 3.46 A dogfish (*Squalus* sp.). The spines present on the dorsal fins are venomous. (Photo: Vidal Haddad Jr.)

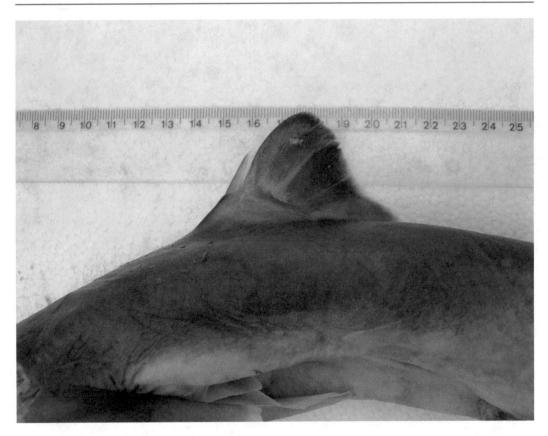

Fig. 3.47 Detail of a venomous spine on a dogfish. (Photo: Vidal Haddad Jr.)

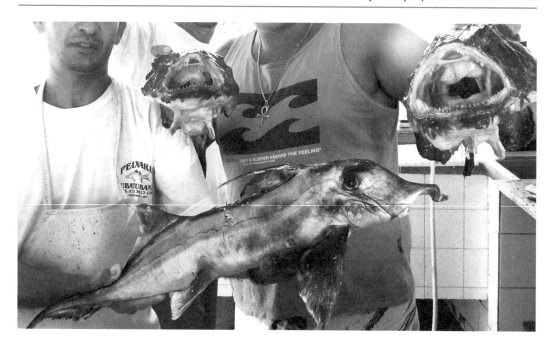

Fig. 3.48 Chimeras are strange fish, related to sharks and rays, and have a venomous spine on their dorsal fin. (Photo: Vidal Haddad Jr.)

Fig. 3.49 Detail of the spine of a chimera. It is very rare for chimeras to injure humans. (Photo: Vidal Haddad Jr.)

The Class Osteichthyes (Bony Fish)

Catfish

The Order Siluriformes: The Families Ariidae, Plotosidae, Ictaluridae, and Pimelodidae

The families of catfish, especially those associated with injuries in humans, are Ariidae and Plotosidae (marine catfish) and Ictaluridae and Pimelodidae (freshwater catfish). Catfish are widely distributed in freshwater and marine environments worldwide (Figs. 3.50, 3.51, and 3.52).

Catfish venom, which is produced by specific glands, is located in three stingers located on the dorsal and pectoral fins (Figs. 3.53 and 3.54). The potency of the venom varies with the species, but all such venom contains proteins that cause intense localized pain, similarly to the effects of acetylcholine and prostaglandin. The wound and surrounding areas show intense pallor and important localized ischemia with occasional skin necrosis [20].

The patient experiences intense localized pain, sweating, and restlessness. There have been reports of fatal evolution, although this was associated with secondary bacterial infections and septicemia rather than with the envenomation per se [20].

Although the vast majority of injuries caused by marine and freshwater catfish occur in individuals who practice fishing activity (amateur and professional fisherman), there is a major problem in coastal areas when small fish are discarded on the sand or in shallow water by fishermen and are then stepped on by walkers in the morning (Figs. 3.55, 3.56, 3.57, 3.58, 3.59, 3.60, 3.61, 3.62, and 3.63a–f). A common complication is breakage of stingers in the wound (Fig. 3.64). Although the fish is dead, the stings may still contain active venom. In some instances, the fish may be rotting, and such injuries can cause severe bacterial infection (Fig. 3.65) [20].

The author has previously reported a fatal injury caused by a catfish on the southeastern coast of Brazil. A fisherman was pulling a net out of a boat when he received an electric shock on the chest from a catfish, and one pectoral sting of the fish pierced his left ventricle (Figs. 3.66 and 3.67). This type of accident can be considered extremely rare because the stings of catfish are

Fig. 3.50 Catfish have three serrated stings on their dorsal and pectoral fins. They are the fish most likely to cause envenomation in humans. (Photo: Vidal Haddad Jr.)

not as large as a stingray's stingers. The coincidence of deep penetration by a stinger and cardiac perforation was exceptional and contributed to the first known description of a human death caused directly by a catfish [21].

Some freshwater catfish also have stingers, which may be venomous or not, in their pectoral and dorsal fins (Fig. 3.63a–f). One study in the Tietê River area in São Paulo State, Brazil, found that injuries involving freshwater catfsh had occurred in 100% of a population of 90 fisherman. The species associated with these injuries was the yellow catfish (*Pimelodus maculatus*) (Figs. 3.68, 3.69, and 3.70). These injuries occurred repeatedly, sometimes even daily, because when the fish were collected in the nets,

the fishermen would break the stingers and discard them on the harbor floor, where they would later be stepped on. Other species of the Pimelodidae family that have stingers and venom are striped and spotted catfish of the genus *Pseudoplatystoma*, and different freshwater catfish around the world, such as the Ictaluridae family in North America, can also cause envenomations [20–25].

The envenomations caused by freshwater and marine catfish are similar to each other and are manifested by pain, swelling, inflammation, frequent secondary infection, and temporary incapacity (Figs. 3.73a–c, 3.74a–c, and 3.75). In this type of injury, it is not unusual to find fragments of stingers retained in the wound [20, 23].

Fig. 3.51 Catfish are common fish, and various species can be captured during the same fishing session. These are four species from the Atlantic Ocean (*from left to right*): a yellow catfish (*Cathorops* sp.), a snake catfish (*Genidens genidens*), a white catfish (*Netuma barba*), and a flag catfish (*Bagre* sp.). (Photo: Vidal Haddad Jr.)

Fig. 3.52 *Plotosus lineatus*, a catfish of the Indo-Pacific, which is responsible for severe human envenomations. (Photo: Vidal Haddad Jr.)

Fig. 3.53 Dorsal stinger of a catfish, showing its serrated appearance. (Photo: Vidal Haddad Jr.)

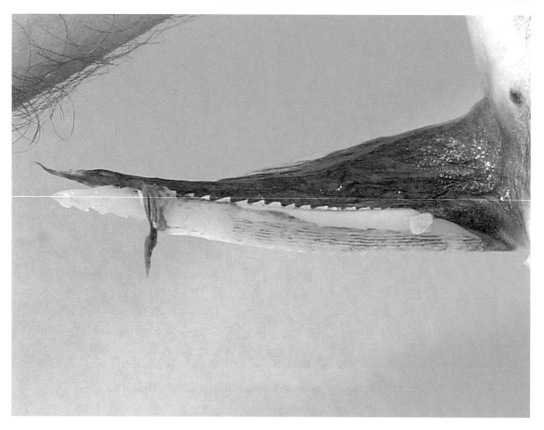

Fig. 3.54 Pectoral stinger of a catfish. The serrated edges can cause severe trauma in addition to envenomation. (Photo: Vidal Haddad Jr.)

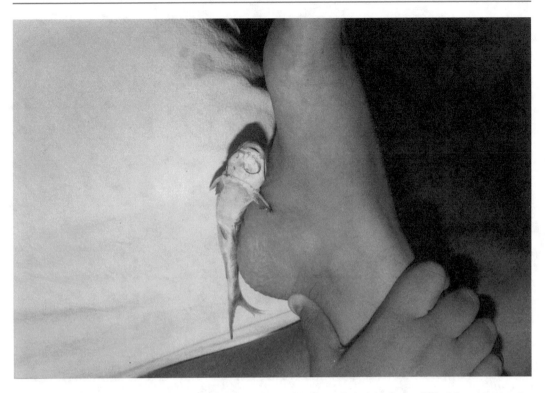

Fig. 3.55 Bathers and walkers can step on small catfish that fishermen discard on the sand. Catfish venom remains active for a certain period after the fish's death. (Photo: Vidal Haddad Jr.)

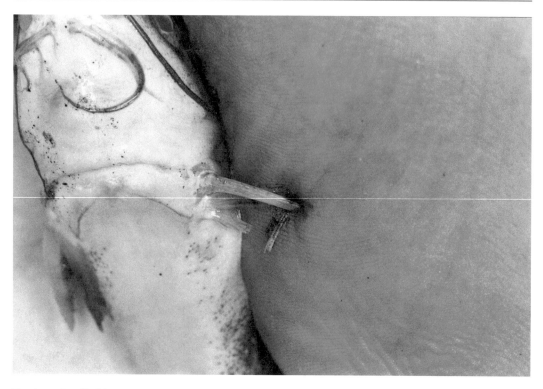

Fig. 3.56   Detail of the anterior image. (Photo: Vidal Haddad Jr.)

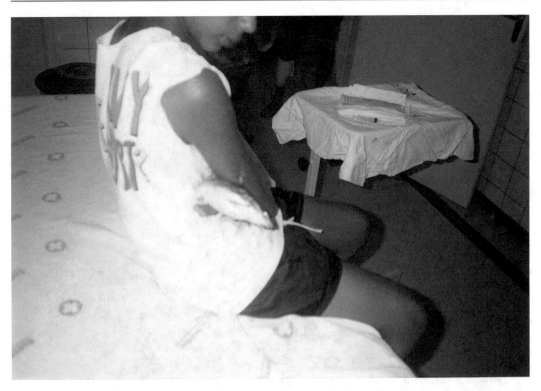

Fig. 3.57 This boy was injured when a friend threw a catfish at him as a joke. Large stingers must always be extracted surgically. (Photo: Vidal Haddad Jr.)

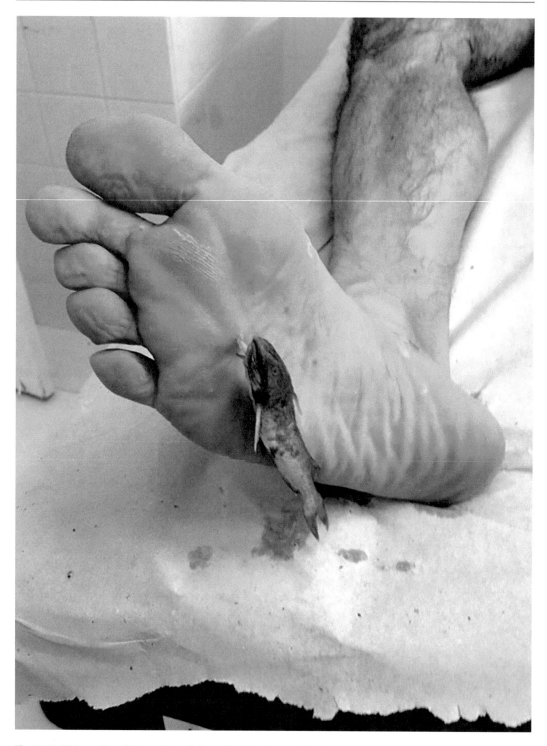

Fig. 3.58 This small catfish was discarded on a beach and stepped on by a bather. (Photo: Vidal Haddad Jr.)

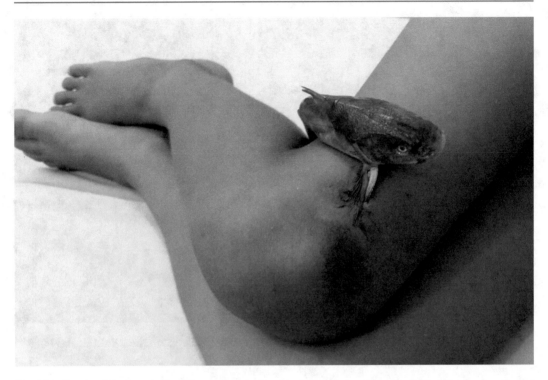

Fig. 3.59 Dead catfish, discarded by fishermen, can be found in shallow water on beaches. This girl was swimming and rolled over the fish on the sand. (Photo: Vidal Haddad Jr.)

Fig. 3.60 A catfsh stinger extracted surgically from a victim after surgery (Fig. 3.59). (Photo: Vidal Haddad Jr.)

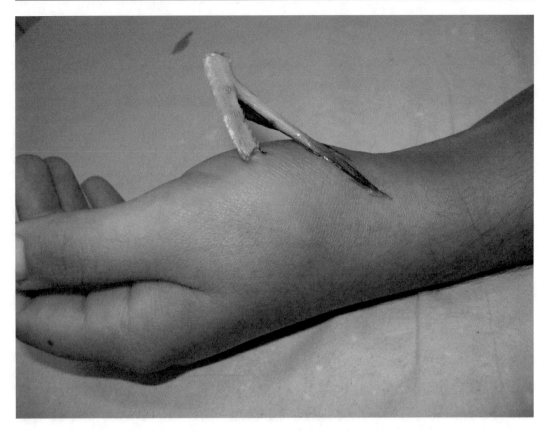

Fig. 3.61 Injuries caused by catfish can occur when cooks are cleaning the fish. In this case, a large pectoral stinger pierced the hand of the victim. (Photo: Vidal Haddad Jr.)

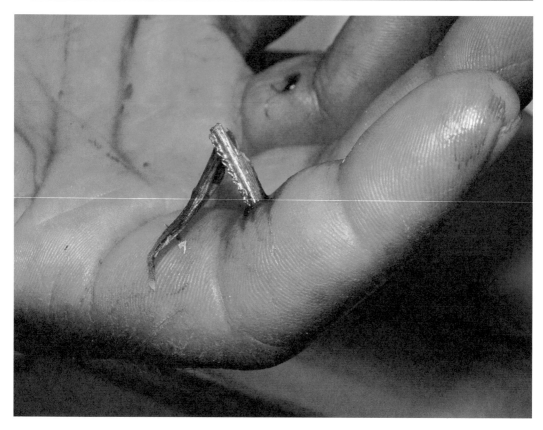

Fig. 3.62 The hand of a fisherman pierced by a catfish stinger is not a rare sight in emergency rooms. (Photo: Vidal Haddad Jr.)

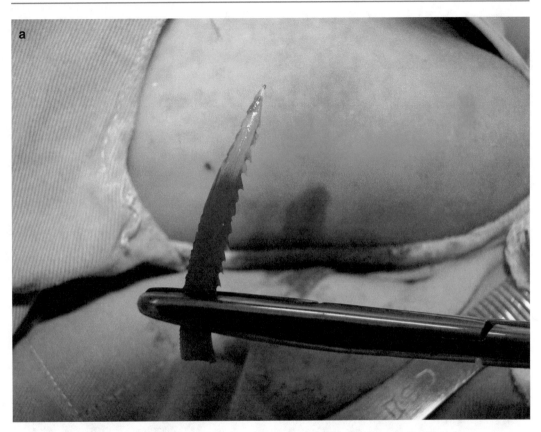

Fig. 3.63 (a) A stinger extracted from the foot of a bather. Note the serrated edges. (Photo: Vidal Haddad Jr.). (b) An injury caused by catfish stinger penetration into a fisherman's retroauricular region, requiring stinger fragment removal. (Photo: Vidal Haddad Jr.). (c) This chronic foot injury in a patient who stepped on a catfish was surgically explored after radiological examination, and a stinger fragment was extracted. (Photo: Vidal Haddad Jr.). (d) An unfortunate combination: a bather and a catfish discarded on the sand by an amateur fisherman (Photo: Vidal Haddad Jr.). (e) This bather had an accident when rolling to the side while lying on a beach. (Photo: Vidal Haddad Jr.). (f) Extraction of catfish stingers can be laborious and requires the stinger entry site to be enlarged with a scalpel. (Photo: Vidal Haddad Jr.)

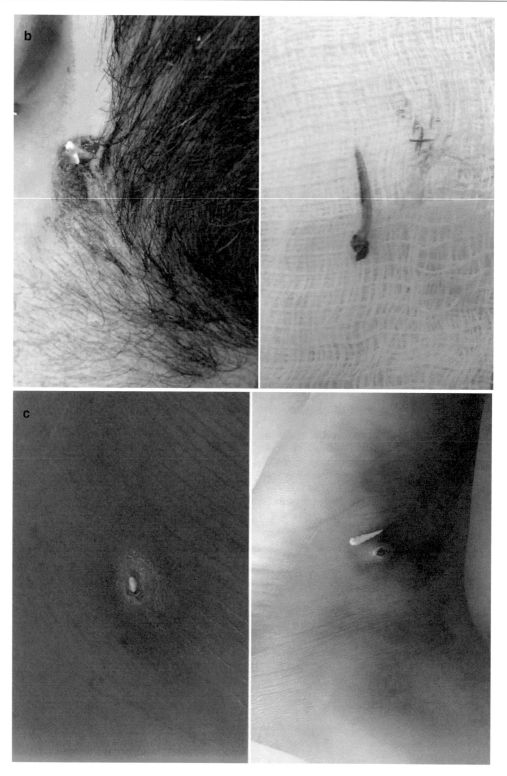

Fig. 3.63 (continued)

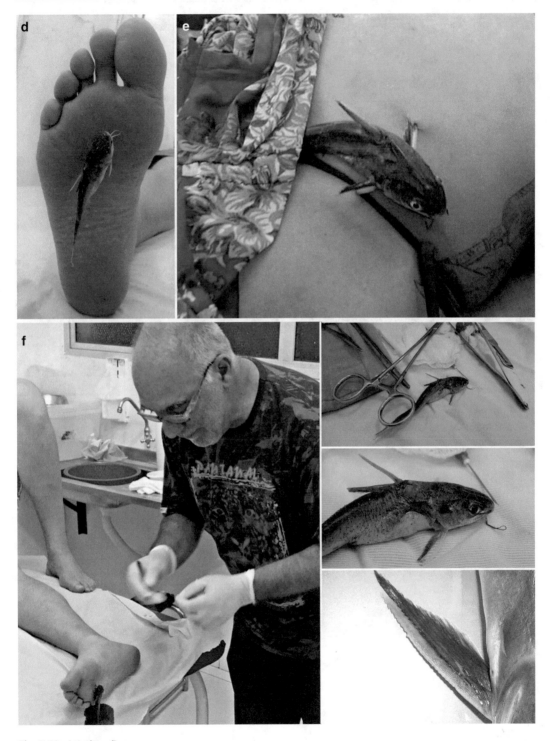

Fig. 3.63 (continued)

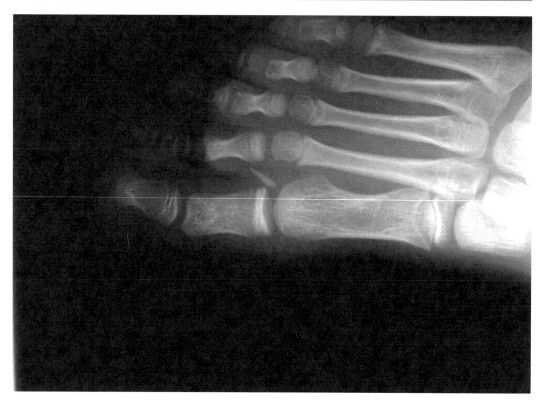

Fig. 3.64 One of the complications of a catfish injury is breakage of the stinger in the wound, which can cause chronic inflammation in the area surrounding the perforation. (Photo: Vidal Haddad Jr.)

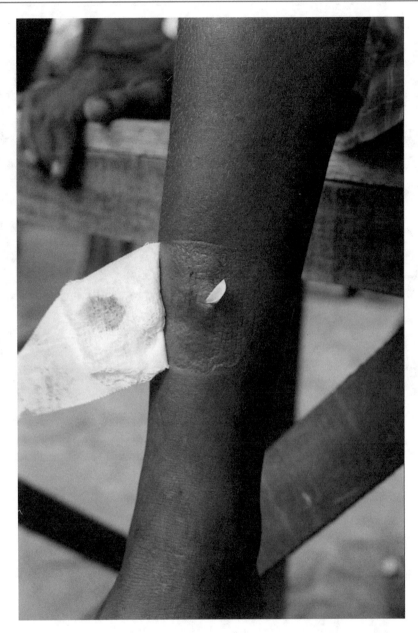

Fig. 3.65 Secondary bacterial infections are a common problem in wounds caused by catfish stingers. This abscess required drainage. (Photo: Vidal Haddad Jr.)

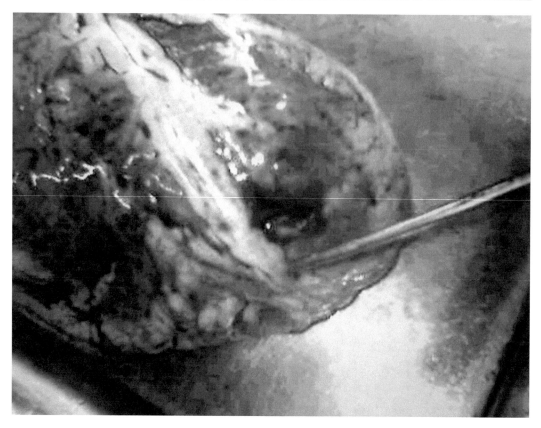

Fig. 3.66 Injuries caused by catfish stingers can occur in unusual places. This fisherman had the left ventricle of his heart perforated while pulling a net containing catfish out of a boat. He bled to death on the beach. (Photo: Vidal Haddad Jr.)

Fig. 3.67 This medium-sized stinger caused the death of a patient in a rare but possible injury. (Photo: Vidal Haddad Jr.)

Fig. 3.68 The yellow catfish (*Pimelodus maculatus*). The venomous stingers of this fish cause the majority of envenomations in South American freshwater environments. (Photo: Vidal Haddad Jr.)

Fig. 3.69 Freshwater catfish have three serrated stingers on their dorsal and pectoral fins. (Photo: Vidal Haddad Jr.)

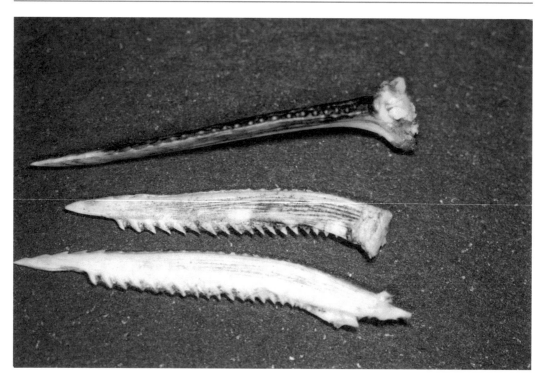

Fig. 3.70   Detail of the bony stingers of a freshwater catfish. (Photo: Vidal Haddad Jr.)

Box 3.2 Catfish

"A 20-year-old man was wounded by a small catfish" while fishing. He then experienced intense pain, edema, and erythema in his right hand for about 6 hours. On the advice of other people on the beach, he applied urine to the wound. Three days after the injury, the site showed severe erythema and swelling, but there was no fever and no purulent secretion (Fig. 3.71). He subsequently sought treatment at a hospital, where detailed clinical and radiological examinations of the sting site were performed (Fig. 3.72).

Comments: Injuries by catfish are the most common envenomations caused by fish (about 80%). In the acute phase of the injury, the fish's toxins cause severe pain and inflammation. Symptoms of painless edema and erythema (which last for a few hours) are consistent with an absence of envenomation. In these cases, it is necessary to look for complications such as retention of stinger fragments and bacterial and fungal infections. If it is suspected that there are foreign bodies at the sting site, radiological examination is necessary.

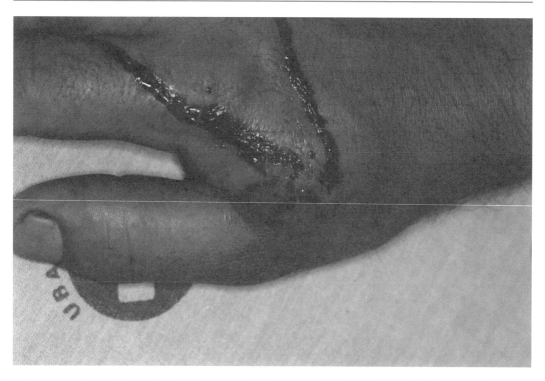

Fig. 3.71 This patient was injured by a catfish while preparing fish in a market. One month after the initial envenomation, he sought medical help for the pain and localized inflammation. (Photo: Vidal Haddad Jr.)

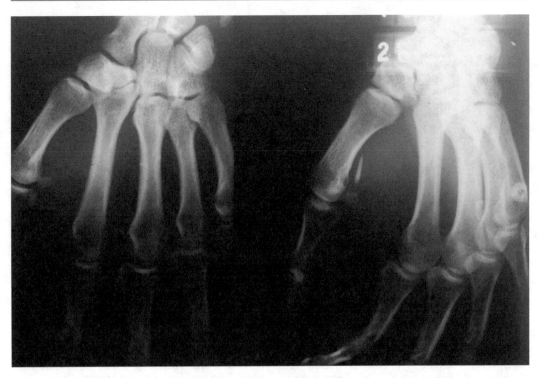

Fig. 3.72 A simple radiological examination of this patient showed that a stinger fragment was responsible for persistence of inflammation at the site of the injury. (Photo: Vidal Haddad Jr.)

Fig. 3.73 (**a**) *Pimelodus maculatus*, a freshwater catfish of South America, is responsible for the majority of the catfish injuries that occur in humans. (Photo: Vidal Haddad Jr.). (**b**) Envenomation with bacterial infection after a sting caused by *Pimelodus maculatus*. (Photo: Vidal Haddad Jr.). (**c**) This kind of occurrence is not uncommon after injuries by catfish, which have small stingers that break easily. (Photo: Vidal Haddad Jr.)

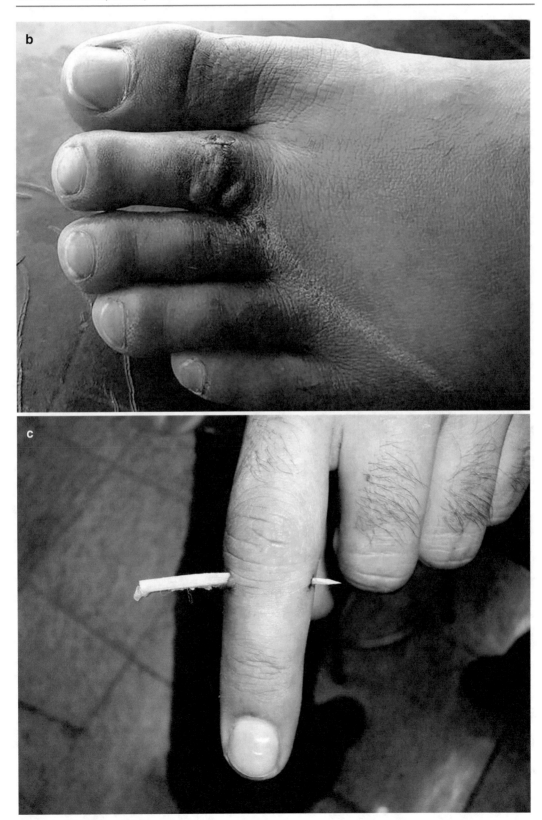

Fig. 3.73 (continued)

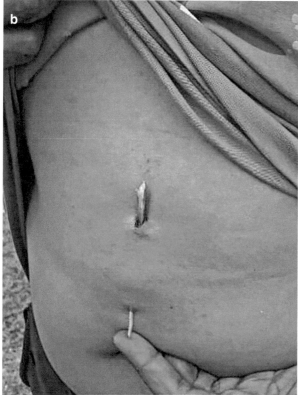

Fig. 3.74 (**a**) The stingers present on the pectoral fins of the surubim (*Pseudoplatystoma* sp.) are the main causes of injuries in humans. (Photo: Vidal Haddad Jr.). (**b**) In addition to causing envenomation, the injuries caused by a surubim result in considerable trauma due to their large size. (**c**) *Pseudoplatystoma fasciatum* is commonly known as the striped catfish or the surubim-cachara. The surubim is a large catfish with venom in its stingers. (Photo: Vidal Haddad Jr.)

Fig. 3.74 (continued)

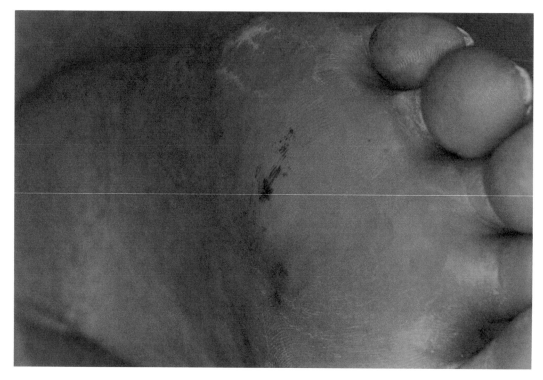

Fig. 3.75 This fisherman stepped on the stinger of a freshwater catfish discarded from a boat and developed erysipelas, manifested by a high fever with intense edema and erythema. (Photo: Vidal Haddad Jr.)

Scorpionfish, Lionfish, and Stonefish

The Families Scorpaenidae and Synanceiidae

The venom of the Scorpaenidae family is similarly composed of thermolabile toxins containing high molecular weight proteins, which cause neurotoxicity and myotoxicity, inducing direct harmful effects on the myocardium.

Lionfish (*Pterois, Dendrochirus,* and *Parapterois*) living in the Indian and Pacific Oceans cause painful injuries but rarely with systemic repercussions (Figs. 3.76 and 3.77) [26]. Envenomations caused by lionfish are not common, even in areas where they are found naturally. However, lionfish do cause sporadic injuries around the world because of the massive trade in them among aquarists. Thus, some breeders have suffered accidents while trying to handle or feed the fish.

Currently, there is an invasion of the species *Pterois volitans* and *Pterois miles* in the Atlantic Ocean, caused by introduction of these species on the eastern coast of the USA and subsequent colonization of the Caribbean and northern South America [26]. Their envenomation provokes severe localized pain and mild systemic manifestations (cardiac and blood pressure effects). The pain is less intense than that caused by scorpionfish, and the risk of death is very low (Figs. 3.78 and 3.79) [26]. There has been one unpublished report of a child's death 3 days after such an injury, but the cause of death was not specified, and it is necessary to evaluate the risk of death from infection by the time elapsed since the envenomation [27].

Scorpionfish (*Scorpaena* spp. and other genera of the Scorpaenidae family) are the most venomous fish in the Atlantic Ocean [28] and are found in all tropical and temperate seas. They can

Fig. 3.76 *Pterois volitans* is the most common species of lionfish in aquariums and has invaded the Atlantic Ocean. This fish was initially restricted to the Indo-Pacific, but its distribution has increased with alarming speed, causing major negative ecological impacts in new areas. (Photo: Vidal Haddad Jr.)

cause very painful injuries accompanied by fever, tachycardia, early adenopathy at the base of the affected limb, and arterial hypotension, but no deaths have been offcially attributed to them. The action of the venom is systemic, unlike the effects of catfish and stingray venoms, which have localized actions [29]. Envenomation occurs through penetration of the victim's skin by the rays on the fins. The rays are grooved and contain venomous glandular tissue (Figs. 3.80, 3.81a–c, 3.82, and 3.83). A massive inoculation (due to perforation by multiple rays) can certainly cause significant systemic symptoms and can potentially be fatal. The dorsal fin of the scorpionfish has 11–17 spines, and the pectoral fins have 11–25 rays (Figs. 3.84 and 3.85) [28, 29].

In the cases previously observed by the author, the victims experienced localized edema and erythema, excruciating pain, incapacity, cold sweats, tachycardia, and diarrhea (Figs. 3.86 and 3.87). One patient suffered behavioral changes and hit his head against the wall of the waiting room; another had hallucinations about an hour after the injury. When the pain was controlled, they regained their self-control and did not seem to remember what they had done. Control of pain after envenomation by scorpionfish is not as simple as that after envenomation by catfish. Sometimes, use of an anesthetic infusion is necessary, or opiates may even be needed. Anyway, application of hot water after the injury is vital. Professional and sports fishermen fishing on

Fig. 3.77 A red lionfish (*Pterois volitans*), showing the sharp tip of one of the rays on its dorsal fin. Envenomation by this fish causes intense pain. (Photo: Vidal Haddad Jr.)

rocky seabeds are particularly exposed to this type of fish, as are divers, because of their curiosity and the uncanny talent these fish have for mimicry, remaining motionless on the seabed or between stones while they lie in wait for prey [4, 5, 28, 29].

The most important fish in the Synanceiidae family are the stonefish (*Synanceia* sp.) and the devilfish, also known as the devil stinger (*Inimicus* sp.) (Figs. 3.94 and 3.95).

Stonefish are the most dangerous fish in the world. They are classified into five species; the most important in terms of the frequency and severity of human injuries are *Synanceia verrucose* (the true stonefish) and *Synanceia horrida* (the estuarine stonefish), which can measure up to 25 cm in length and live in the Indo-Pacific region. They feed on small fish, which they capture through fast mouth movements. Otherwise, stonefish move slowly, and their defense strategy

is to stay semiburied in sand under the water and change their coloration to blend in with the substrate. Additionally, they can aggregate mud and other materials in their body to produce a sticky secretion [1–9].

Stonefish have venomous rays on their fins, especially on the dorsal fin (Fig. 3.96). The genus *Synanceia* possesses 13 sharp venomous spines along its dorsal fin. These rays on the fins contain adapted venom glands covered by a warty sheath, and they can inject 5–10 mg of venom per spine [1]. When the fish is stepped on, the sheath squeezes venom out of the the glands. The most common victims are divers, tourists, and fishermen. The spines can penetrate rubber footwear, and it is difficult to prevent injuries. Even after death, stonefish lying on the sand can cause envenomations for nearly 24 hours.

The venom of stonefish has neurotoxic, myotoxic, cardiotoxic, and cytotoxic effects. The

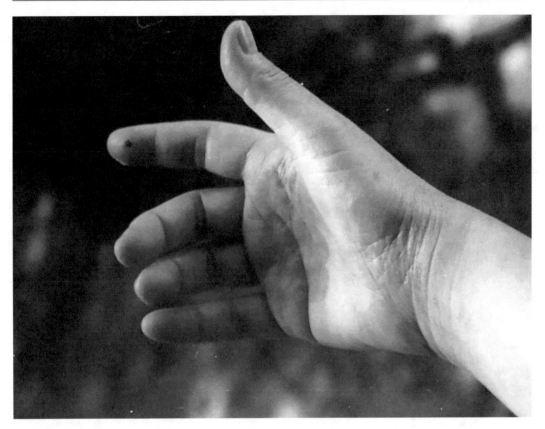

Fig. 3.78 A single perforation on the hand of this aquarist caused excruciating pain, malaise, and localized edema and erythema, but there were no systemic manifestations (other than symptoms associated with the pain). (Photo: Vidal Haddad Jr.)

main symptom of envenomation is violent pain, which is associated with P-substance receptors. A possible explanation is that the venom may contain toxins similar to endogenous transmitters that provoke pain sensations and localized inflammation [4, 5]

When death is caused by envenomation, it occurs during the initial hours after the injury. The victim suffers an initial risk of drowning because of the pain. In the latter phases, bacterial infections can increase the risk of problems. Various signs and symptoms (nausea/vomiting, cardiac arrhythmias, syncope, delirium, and seizures) can be associated with the intense pain, which can provoke shock. Lehmann and Hardy [30] have previously described a case of pulmonary edema, which responded to antivenom therapy, following stonefish envenomation caused by seven spines.

First aid for stonefish stings involves copious washing of the wound site, removal of spines that may be present in the wound, immersion of the site of the injury in tolerably hot water (about 45° C) for 30–90 minutes in an attempt to control the pain, and prompt transport to an emergency medical center.

The components of the fish's venom are thermolabile. This fact surely has importance in the action of hot temperatures on the wound but is not the only factor in its control; the venom of Scorpaenidae (and that of other fish) provokes marked vasoconstriction in the area of the wound, and the consequences are ischemia of the tissues, excruciating pain, and localized blanching and erythema [11, 31]. Hot water promotes vasodilatation and vascular permeability, decreasing ischemia and pain. When the patient takes the affected area out of the hot

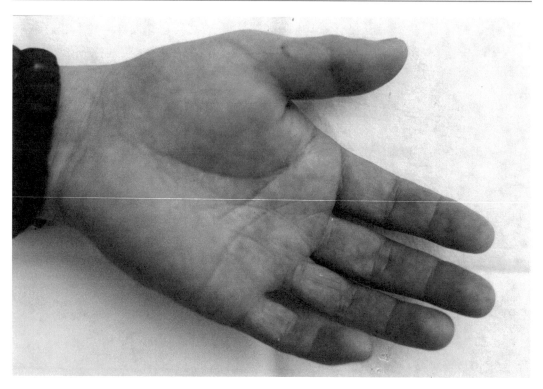

Fig. 3.79 This patient suffered an envenomation with predominance of pain and localized phenomena. (Photo: Vidal Haddad Jr.)

Fig. 3.80 Scorpionfish are the most venomous fish in the Atlantic Ocean. Their venom has systemic effects and can cause cardiological and neurological symptoms. (Photo: Vidal Haddad Jr.)

a

Fig. 3.81 (**a**) The rays on the fins (especially the dorsal fin) of the scorpionfish have grooves containing venomous tissue. (Photo: Vidal Haddad Jr.). (**b**) *Scorpaena plumieri* is the most common scorpionfish in the Atlantic Ocean and the Americas. (Photo: Vidal Haddad Jr.). (**c**) The red scorpionfish (*Scorpaena scrofa*) is the largest eastern Atlantic scorpionfish. (Photo: Vidal Haddad Jr.)

Fig. 3.81 (continued)

Fig. 3.82 *Scorpaena plumieri* is commonly known as the spotted or black scorpionfish. The injuries caused by this fish can be severe, causing systemic disease. (Photo: Vidal Haddad Jr.)

Fig. 3.83 The red scorpionfish, or barbfish (*Scorpaena brasiliensis*), is a common species in the western Atlantic Ocean. This species is found in shrimp fishermen's nets, causing various envenomations

Fig. 3.84 The rays on the dorsal fin of a spotted scorpionfish show its danger to anyone who touches the fish inadvertently. (Photo: Vidal Haddad Jr.)

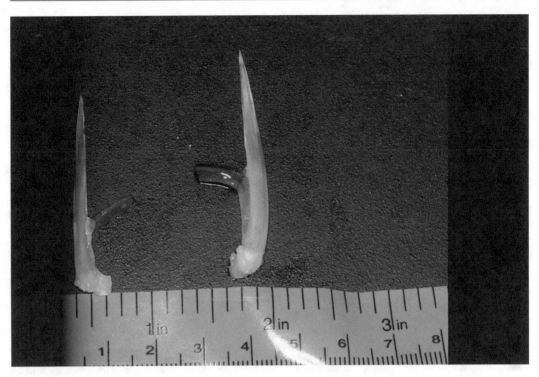

Fig. 3.85 Rays on the dorsal fin of *Scorpaena plumieri*, showing the venomous glandular tissue present in the grooves. (Photo: Vidal Haddad Jr.)

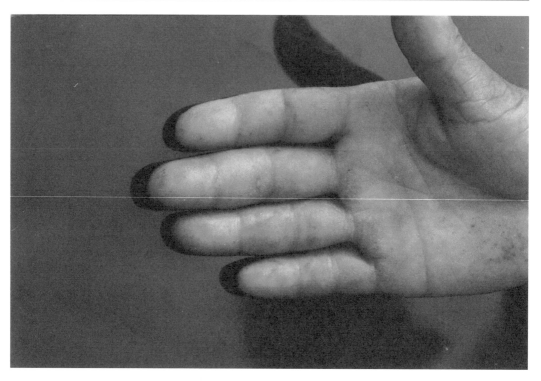

Fig. 3.86 Edema, erythema, and impaired mobility of the hand in a fisherman injured by a spotted scorpionfish. This photo was taken 1 month after the injury. In the acute phase, he suffered severe pain, localized inflammatory phenomena, and systemic symptoms. (Photo: Vidal Haddad Jr.)

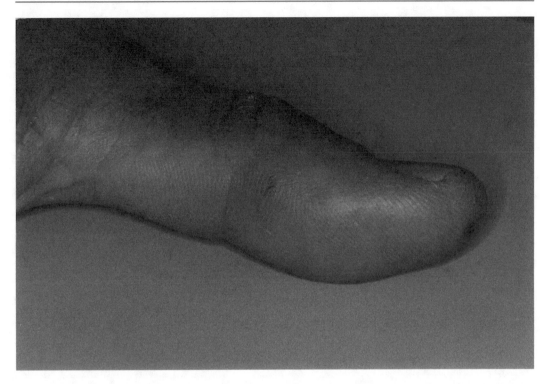

Fig. 3.87 A perforation caused by a scorpionfish on the finger of this fisherman provoked intense pain and malaise for about 24 hours. (Photo: Vidal Haddad Jr.)

Box 3.3 Scorpionfish

A 28-year-old man, who was a sport fishing practitioner, reported having caught a dark-colored fish about 40 cm long at a coastal beach in Ubatuba (São Paulo State, Brazil). The fish raised its spines as it was removed from the hook; as a result, the fisherman carelessly dropped it on his left foot, which was penetrated by three spines on the dorsal fin (Fig. 3.88). The fish was subsequently identified as a black scorpionfish (*Scorpaena plumieri*; see Fig. 3.89). Immediately after the injury, the victim started to feel localized burning and tingling in his left hallux. After he had walked for about 100 m to his car, the pain was unbearable, radiating to his leg, thigh, and abdomen.

During his transportation to a hospital, he experienced strong tremors, blurred vision, lack of salivation, dysarthria, intense coald sweats, loss of spatial and temporal location, and excruciating pain. He was given spinal anesthesia and systemic opioids, but they had no effect on the pain. After 3 hours, the systemic phenomena disappeared, but localized erythema occurred. It partially regressed when the patient immersed his feet in hot water for about two 2 hours. The swelling persisted for about 7 days after the envenomation. At that time, the patient still had slight tremors.

Comments: Injuries by scorpionfish are serious. The venom has systemic effects and can be lethal in cases of a massive inoculation. The clinical manifestations are severe and differ from the localized manifestations caused by other fish, such as catfish and stingrays. There is no antivenom to treat scorpionfish envenomation; thus, only the consequent symptoms can be treated.

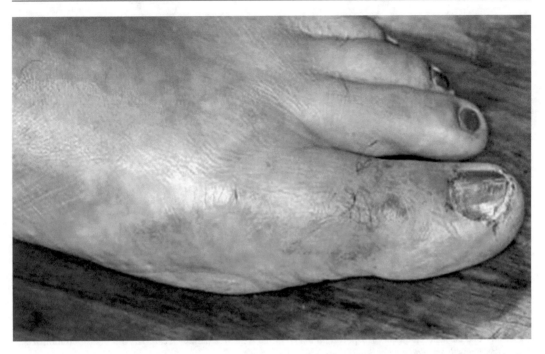

Fig. 3.88 Perforations caused by three rays on the fins of a scorpionfish, which caused severe systemic effects after envenomation. (Photo: Vidal Haddad Jr.)

Fig. 3.89 The black scorpionfish (*Scorpaena plumieri*) of the western Atlantic Ocean. The different scorpionfish species in the genus *Scorpaena* are similar. (Photo: Vidal Haddad Jr.)

Box 3.4 Lionfish

A 37-year-old man suffered an injury while handling a lionfish in a home aquarium (Fig. 3.90). When it was examined, there were two punctures on the first finger of his right hand, caused by the rays on the dorsal fin of the fish. The patient immediately experienced intense pain, followed by localized edema and erythema, which spread to the entire hand (Fig. 3.91). He was advised to immerse the hand in hot water, which reduced the pain. The next day, he noted a blister filled with clear fluid in the area of the injury. The pain disappeared within about 12 hours, but localized inflammation persisted at the site after 5 days.

Comments: Envenomations by lionfish can cause systemic phenomena, but usually they cause only localized signs and symptoms, which are intense. Inflammation may be noted, followed by blistering. The treatment is symptomatic, with an emphasis on pain control (using hot water) and prevention of secondary infections.

Fig. 3.90 The lionfish species that is most commonly kept in aquariums and is predominant in the invasion of the Atlantic Ocean is *Pterois volitans*. (Photo: Vidal Haddad Jr.)

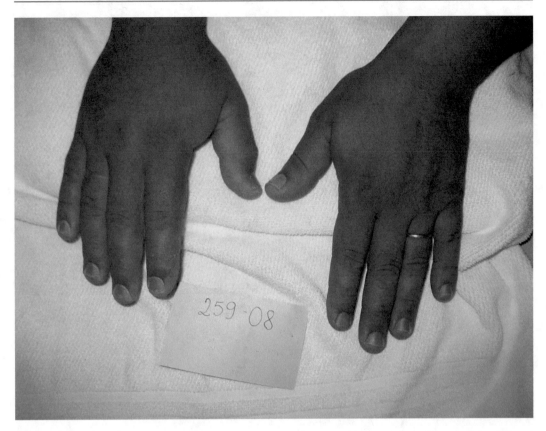

Fig. 3.91 This patient was handling a lionfish in a home aquarium when he was wounded on the first finger of his right hand by the rays on the dorsal fin of the fish. The patient immediately experienced intense pain, followed by localized edema and erythema, which spread to the entire hand. (Photo: Vidal Haddad Jr.)

Box 3.5 Stonefish

A 42-year-old man was observed at an emergency medical center in Brazil after suffering a perforation in the medial part of the second finger of his hand while handling a stonefish in an aquarium (Fig. 3.92). He experienced excruciating localized pain, dizziness, and malaise. The clinical examination showed a puncture and mild edema/erythema (Fig. 3.93). The patient had no systemic manifestations. He was treated with hot water and painkillers, and experienced gradual improvements in his signs and symptoms, becoming asymptomatic after 2 days and being discharged after 3 days.

Comments: Stonefish have been proved to cause human deaths, which have been associated with the envenomation itself rather than being due to complications such as disseminated bacterial infections. The extreme severity of the manifestations is related to the venom's powerful cardiovascular action, ability to increase vascular permeability, and systemic action. It is not coincidental that this is the only type of envenomation by fish that warrants regular production of antivenom serum, although deaths from stonefish envenomation are rare.

Fig. 3.92 The estuarine stonefish (*Synanceia horrida*) can cause severe human envenomation, which can be fatal. (Photo: Vidal Haddad Jr.)

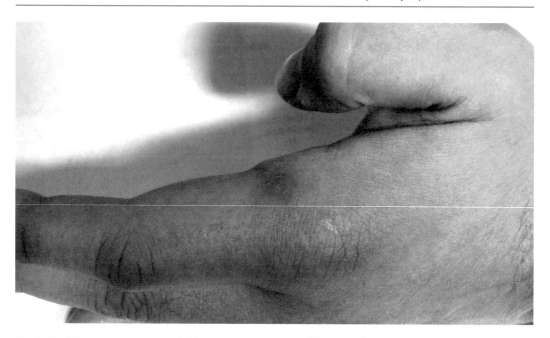

Fig. 3.93 This aquarist was wounded by an estuarine stonefish in a domestic aquarium in Brazil. An isolated perforation of his finger by a ray on one of the fish's fins caused intense inflammation and pain but no systemic manifestations. (Photo: Vidal Haddad Jr.)

Fig. 3.94 The true stonefish (*Synanceia verrucosa*) is one of the most venomous fish in the world. Its envenomation can cause human deaths. (Photo: v Haddad Jr.)

Fig. 3.95 The devilfish is part of the Synanceiidae family and, as a stonefish, can cause serious envenomations. (Photo: Vidal Haddad Jr.)

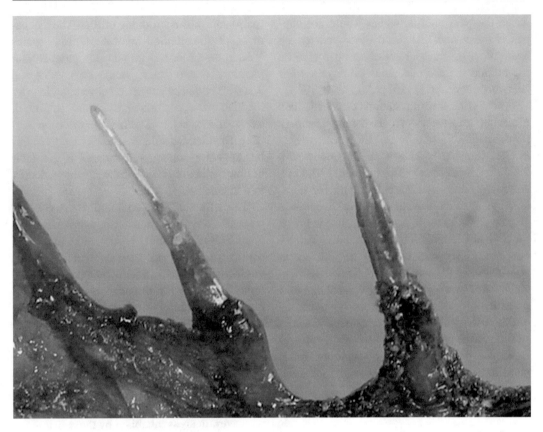

Fig. 3.96 The rays on the fins of the stonefish have grooves containing venomous glandular tissue. (Photo: Vidal Haddad Jr.)

water, the pain returns, which would not happen if the venom had been destroyed by the heat. Local or regional anesthesia can also be useful, and the use of tetanus prophylaxis has been recommended [11, 31].

Severe clinical manifestations can necessitate lifesaving measures, such as cardiopulmonary resuscitation.

CSL (previously known as Commonwealth Serum Laboratories) (Melbourne, Victoria, Australia) produces a specific serum for intramuscular use: the recommended doses are one ampule (2000 U) for 1–2 punctures, two ampules for 3–4 punctures, and three ampules for >4 punctures. The serum should be diluted in 50–100 mL of isotonic sodium chloride solution and infused over a period of at least 20 minutes. As with all hyperimmune antisera of equine origin, there are risks of allergic reactions and serum sickness in the patient. Australian sources recommend use of subcutaneous epinephrine, intramuscents with known hypersensitivity, addition of an intramuscular corticosteroid.

The Family Batrachoididae

Toadfish

Toadfish are fish of the Batrachoididae family, and the *Batrachoides*, *Porichthys*, and *Thalassophryne* genera are present in tropical and temperate marine and estuarine waters throughout the world (Figs. 3.97 and 3.98a, b). The venom of toadfish provokes neurotoxicity (pain) and skin-necrotizing effects. The mechanism of inoculation involves hollow spines located on the dorsal fin and in a preopercular position (one on each side) (Figs. 3.99 and 3.100). The initial studies on envenomations by toadfish were performed by Froes in Brazil, using the species *Thalassophryne nattereri* [32]. Toadfish have the most developed venomous apparatus among venomous fish, and their hollow spines can inject their venom deeply into the victim.

The effects of the venom are localized, and there are no systemic manifestations of envenomation. This type of injury is very common in some regions of the world, such as in northern and northeastern regions, where they can be easily found in marine and estuarine waters. The injuries are very painful, often cause mild to moderate localized necrosis, and can make the victim incapable of work for days (Figs. 3.101 and 3.102) [33, 34].

Use of hot water relieves the pain, but the swelling and inflammation persist, causing these fish to be greatly feared by fishermen and others who work in areas inhabited by these fish.

The author has previously observed an envenomation caused by *Porichthys porosissimus* (a toadfish species that was recently proved to be venomous) in a fisherman, with penetration of a dorsal spine into the right index finger, causing instantaneous and pulsating pain, which lasted for about 2 hours and then subsided spontaneously without use of any medication [33].

Fig. 3.97 *Thalassophryne nattereri* is a venomous fish of the Atlantic Ocean. This toadfish causes envenomations in estuarine areas. (Photo: Vidal Haddad Jr.)

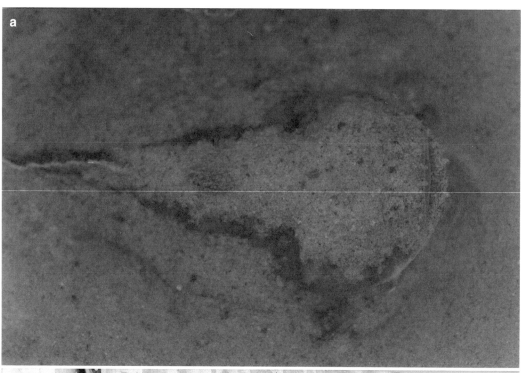

Fig. 3.98 (**a**) The toadfish can stay semiburied in sand or mud, where it may be stepped on by its victim. Envenomation is caused by hollow dorsal spines or, more rarely, by preopercular spines. (Photo: Vidal Haddad Jr.) (**b**) The niquim or Atlantic toadfish is causer of the majority of envenomations caused by venomous fish in Northeast region of Brazil They have the most sophisticated venom apparatus of the all fish. (Photo: Vidal Haddad Junior)

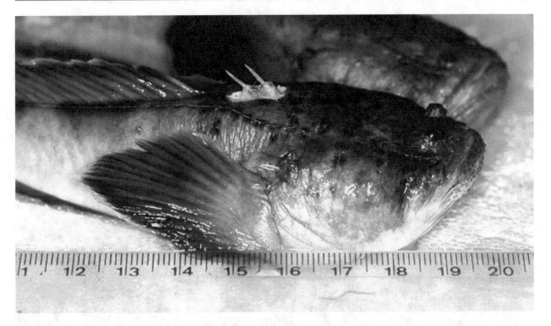

Fig. 3.99 The dorsal spines of the toadfish are linked to true venom glands. (Photo: Vidal Haddad Jr.)

Fig. 3.100   Detail of the dorsal spines and glands. (Photo: Vidal Haddad Jr.)

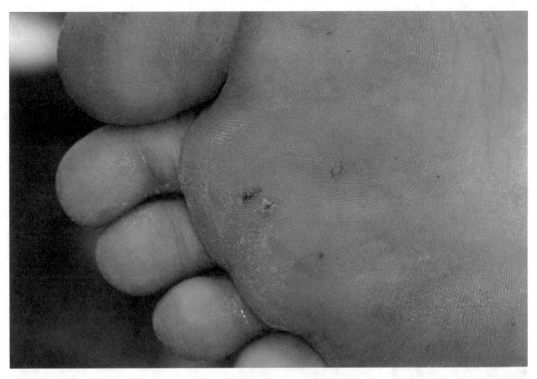

Fig. 3.101 Injuries caused by the dorsal spines of a toadfish. This fisherman suffered intense pain and inflammation but no skin necrosis. Note the double perforation. (Photo: Vidal Haddad Jr.)

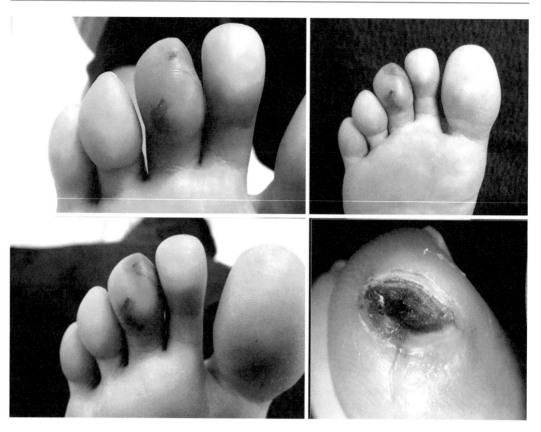

Fig. 3.102 This patient was wounded by a *Thalassophryne nattereri* toadfish in a lagoon, and an acute process involving edema, erythema, and pain evolved into localized necrosis, with subsequent development of mild skin necrosis. (Photo: Vidal Haddad Jr.)

The Family Acanthuridae

Surgeonfish

The fish of the Acanthuridae family have two sharp and pointy blades on both sides of their tail (Figs. 3.103 and 3.104). These blades can cause incised wounds similarly to surgical blades, with intense bleeding and pain that are always disproportionate to the injury. These features denote a probable toxic effect, although this has not yet been studied (Fig. 3.105). Surgeonfish are common on reefs and include various species with the common feature of caudal blades [4, 5].

Fig. 3.103 The fish of the Acanthuridae family have two sharp blades on their tail, which can cause severe traumatic lesions in humans. (Photo: Vidal Haddad Jr.)

Fig. 3.104 Detail of a sharp blade on a surgeonfish. (Photo: Vidal Haddad Jr.)

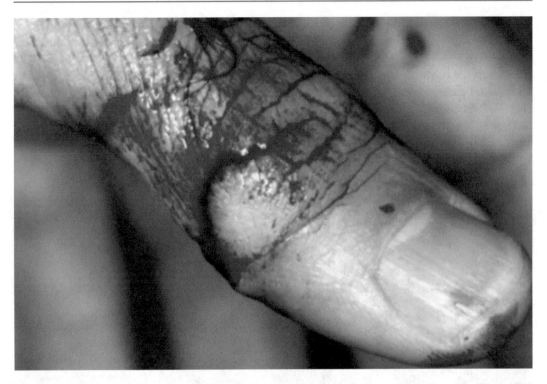

Fig. 3.105 This incised lesion with intense bleeding is characteristic of lesions caused by surgeonfish. (Photo: Vidal Haddad Jr.)

The Family Muraenidae

Moray Eels

Moray eels are fish with a serpentiform appearance and belong to the Muraenidae family (Figs. 3.106 and 3.107). These fish have pointed teeth capable of causing severe lacerations (Fig. 3.110). The pain is disproportionate to the wounds and tends to persist for 12 hours. This fact has caused suspicion that glands present on the palate of the fish can produce venom. Sand moray eels (*Gymnothorax ocellatus*) are found in shrimp fishing nets, and their bite is feared by fishermen. The pain is relieved by hot water. Other larger species of moray eel are often found by divers in deep waters and can cause injuries in people who venture close to these fish [4, 5, 35].

Fig. 3.106 Moray eels have sharp teeth, and their bites cause extensive lacerated lesions. (Photo: Vidal Haddad Jr.)

Fig. 3.107 Detail of the teeth of a moray eel. (Photo: Vidal Haddad Jr.)

Box 3.6 Weeverfish

A 38-year-old woman suffered a weeverfish sting when she stepped on something sharp and pointed while walking in murky knee-deep water on a sandy beach in the Algarve, Portugal (Figs. 3.108 and 3.109). Immediately after the injury, she felt intense burning pain, radiating to the root of the affected limb. Over the course of about 1 hour, the pain gradually increased until it became unbearable, at which time, the victim was referred to an emergency medical center. Over the following hours, the site became erythematous and edematous, with edema progressing to the ankle. The patient had low-grade fever, cold sweats, chills, nausea, and arterial hypotension, which were attributed to her agitation and her pain. She was treated for her symptoms and discharged the next morning.

Comments: Weeverfish are venomous fish found on European coasts and in North Africa (Fig. 3.109). They cause injuries in fishermen and bathers when they are handled or stepped on. Their envenomation causes intense inflammation, pain, systemic phenomena linked to the pain, and malaise. Serious infectious processes can occur, aggravating the injury. Treatment is symptomatic, with control of pain and inflammation.

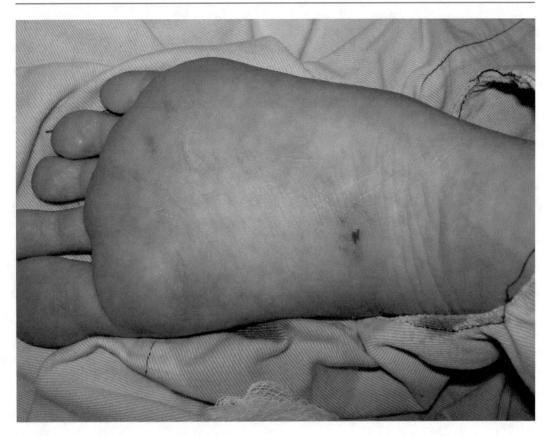

Fig. 3.108 This puncture on the foot of a bather, caused by a weeverfish, caused intense pain and malaise. (Photo: Vidal Haddad Jr.)

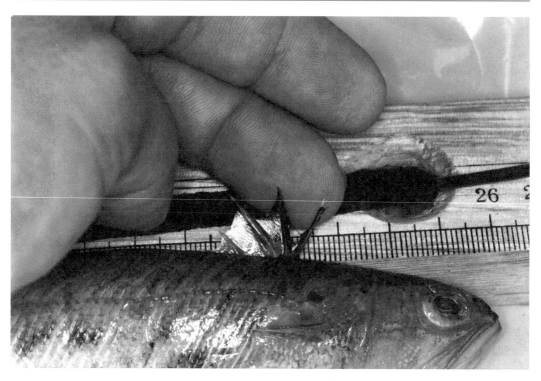

Fig. 3.109 Weeverfish have a very specialized dorsal and preopercular apparatus for envenomation. (Photo: Vidal Haddad Jr.)

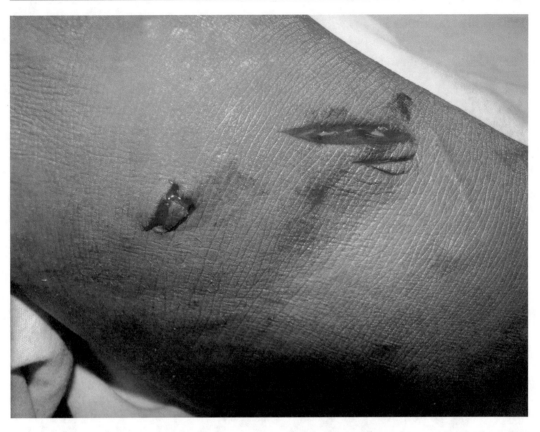

Fig. 3.110 The lesions caused by moray eel bites are irregular because of the position of the teeth in the fish's mouth. (Photo: Vidal Haddad Jr.)

The Family Trachinidae

Weeverfish

Trachinidae have proteinaceous venom, which is thermolabile and induces skin necrosis (Figs. 3.111, 3.112 and 3.113). The majority of injuries cause localized manifestations, with intense pain and occasional necrosis. Weeverfish are venomous fish that cause the greatest number of envenomations by fish in Europe, especially in Atlantic and Mediterranean waters. In the Atlantic, they are found only in the waters around Europe and North Africa [4, 5, 7, 9].

Fig. 3.111   Different species of *Trachinus* in a fish market in Malaga, Spain. (Photo: Vidal Haddad Jr.)

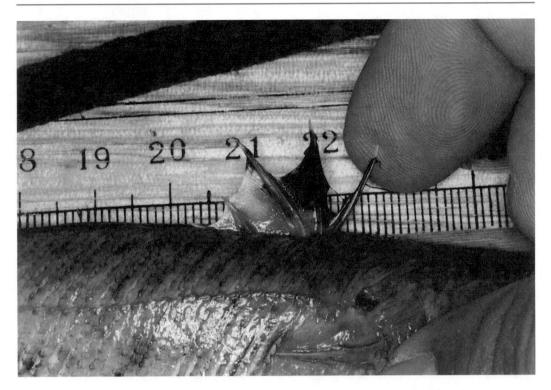

Fig. 3.112 The weeverfish has dorsal and preopercular venomous spines. (Photo: Vidal Haddad Jr.)

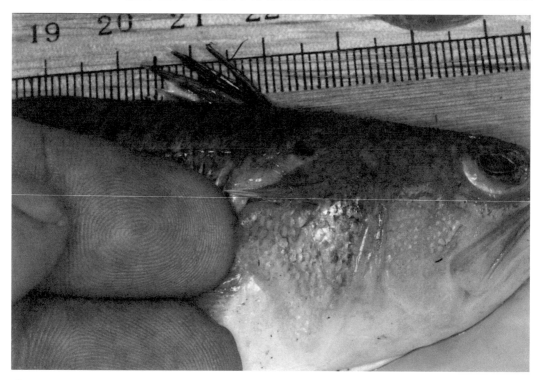

Fig. 3.113 Weeverfish present venomous opercular spines. (Photo: Vidal Haddad Jr.)

Other Venomous Fish

In addition to the venomous fish that are most commonly associated with human injuries, there are other fish that cause sporadic injuries. In some of them, the presence of toxins has not yet been proven, although the victims manifest prolonged pain and localized inflammatory phenomena typical of envenomations. Among these fish, squirrelfish (*Holocentrus adscensionis* and other species) have spines in a preopercular position, which have been described as being covered with a toxic glandular epithelium (Fig. 3.114a). The Carangidae family includes two main species that are considered venomous: the barred grunt (*Conodon nobilis*) and the Castin leatherjacket or guaivira (*Oligoplites saliens*), with sharp dorsal and anal spines that are presumed to be coated with venom glands (Fig. 3.114b). The flying gurnard (*Dactylopterus volitans*) and the sea robin (*Prionotus* sp.) have sharp cephalic spikes, which are also presumed to be covered with venomous glands (Figs. 3.115 and 3.116). The Siganidae family includes venomous species (Fig. 3.117). Stargazers, which are part of the Uranoscopidae family (*Astroscopus* spp.) have a flat body, stay semiburied in sand or mud under the water, and have glands in a preopercular position that probably produce venom (Fig. 3.118). The anglerfish or batfish (*Lophius* spp.) has a bizarre appearance and venomous spines on its body (Fig. 3.119) [4, 5, 7, 9].

Fig. 3.114   (a) The squirrelfish (*Holocentrus* sp.) (b) A leatherjacket, also known as a guaivira (*Oligoplites* sp.)

Figs. 3.115 The flying gurnard (*Dactylopterus volitans*)

Figs. 3.116 The sea robin (*Prionotus* sp.) have sharp cephalic spikes

Fig. 3.117 The fish of the Siganidae family are venomous fish with toxins in the rays on their fins. (Photo: Vidal Haddad Jr.)

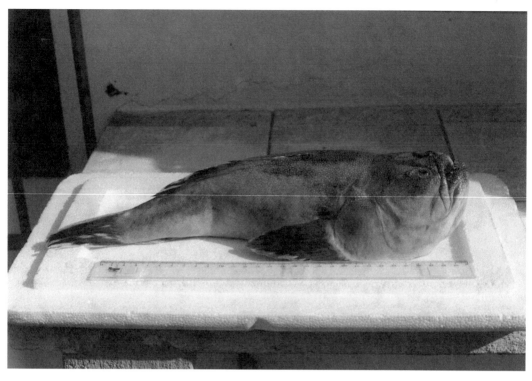

Fig. 3.118 Stargazers have preopercular venomous spines and deliver small electric shocks to defend themselves. (Photo: Vidal Haddad Jr.)

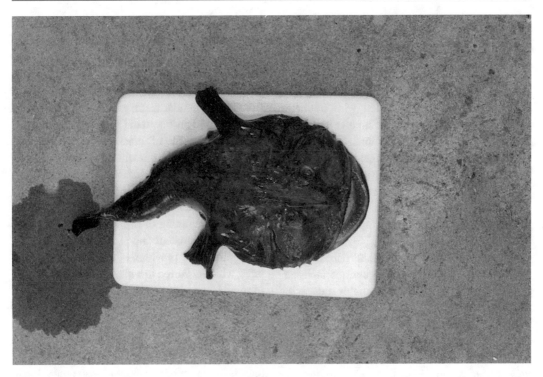

Fig. 3.119 The monkfish, anglerfish, tamboril or batfish have body spines that are considered venomous. (Photo: Vidal Haddad Jr.)

Traumatogenic Fish

Marine Fish

Although any fish can cause injury to humans through spines, stingers, and teeth, some species are more frequently associated with traumatic injuries. These cases may not involve envenomation, but there may be lacerations, severe bleeding, and late bacterial and fungal infections. The fish responsible for these injuries include barracudas (the genus *Sphyraena*), needlefish (Belonidae family), swordfish (*Trichiuris lepturus*), and triggerfish (*Balistes* sp.) (Figs. 3.120, 3.121a, b, and 3.122a, b) [4, 5, 36].

Marine environments contain fish of large sizes, which can cause severe injuries in fishermen and divers. The mechanisms of aggression are collisions with the bodies of the fish, bites, and wounds caused by spines and stingers. This group, which comprises groupers and snappers (the *Epinephelus*, *Lutjanus*, and *Mycteroperca* genera), includes the jewfish, or Goliath grouper (*Epinephelus itajara*, a giant weighing up to 450 kg), and the snook (*Centropomus* sp.), which has sharp blades near both opercula (Figs. 3.123 and 3.124) [4, 5]. Other potentially injurious fish are swordfish (the genus *Xiphias*), sailfish (the genus *Istiophorus*), blue marlin (*Makaira nigricans*), and white marlin (*Kajikia albida*, previously known as *Tetrapturus albidus*). These fish possess a highly traumatogenic "beak," which is used for capturing prey and for defense (Figs. 3.125 and 3.126) [37].

Freshwater Fish

Piranhas are the main fish associated with traumatic lesions in freshwater environments (Fig. 3.127), but, despite the folklore surrounding them, there have been no documented attacks on humans by shoals of piranhas, and there have been only a few reports of large animals being devoured by them. In truth, piranhas are voracious carnivores that act as decomposers in nature. They are strongly attracted by blood in the water and agitated movements by the victim, but attacks by shoals are rare. In a case series of humans who were supposedly devoured by piranhas, their necropsies showed that the victims had actually died from drowning or from cardiovascular disease, and that their piranha bites occurred postmortem [38].

The real profile of attacks by piranhas shows single deep bites in humans. The bite is oval or rounded, and causes laceration and bleeding (piecemeal injuries) (Figs. 3.128 and 3.129a–c). These bites occur in ponds and in water bodies formed by damming of rivers, and they occasionally occur in amateur and professional fishermen. The author has previously observed nearly 320 bites, which occurred in bathers on beaches formed in small tributaries of dammed rivers in Brazil. All of the bites occurred on the extremities (especially on the heels, legs, hands, and fingers), and the responsible species was the piranha *Serrasalmus maculatus* (Fig. 3.130) [39, 40]. This profile of bites is the cause of the piranha's bad reputation in the rivers of South America—not attacks by schools of fish on humans, which, although not impossible, are extremely rare. A massive attack would require a school of fish (which is more likely to be associated with *Pygocentrus nattereri*), agitation and/or blood in the water, and other factors combined. No accidents of this type have been described in the literature.

Various other attacks with this same profile have been reported in similar areas in Brazil and elsewhere in South America. The probable causes are the male piranha's defense response to postures performed in shallow water in a dam, combined with intense agitation of the water by sometimes thousands of bathers in a small space and lack of the piranha's predators in the area. Control of piranha biting was made possible by installation of nets a few meters from the beach and, in the intermediate term, reintroduction of dorado, or golden dorado (*Salminus brasiliensis*), a natural predator of piranhas [39, 40].

Fig. 3.120 Barracudas are large fish and can be aggressive toward humans, but attacks by them are rare. (Photo: Vidal Haddad Jr.)

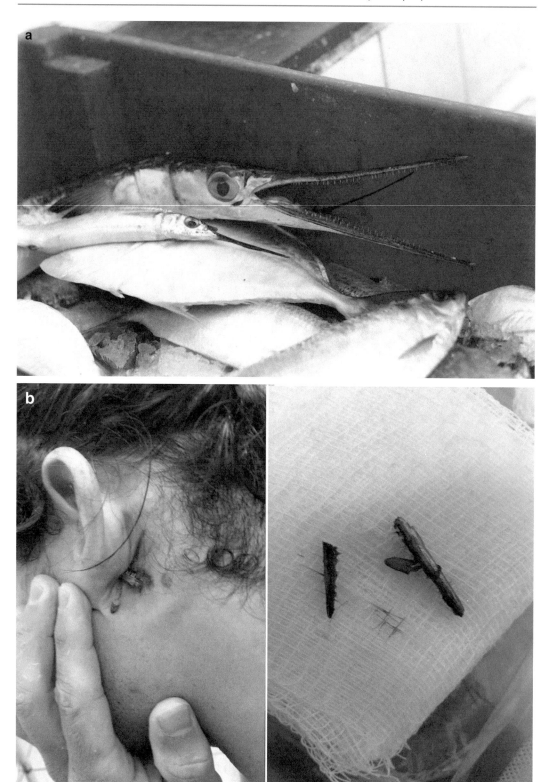

Fig. 3.121 (**a**) Needlefish (which are part of the Belonidae family) have sharp teeth. They can attack lights on boats and cause lesions in fishermen. (Photo: Vidal Haddad Jr.). (**b**) Shoals of needlefish can collide with surf-ers and swimmers, causing injuries that can be serious, depending on where they penetrate the body and whether the fish's "beak" breaks off in the wound. (Photo: Vidal Haddad Jr.)

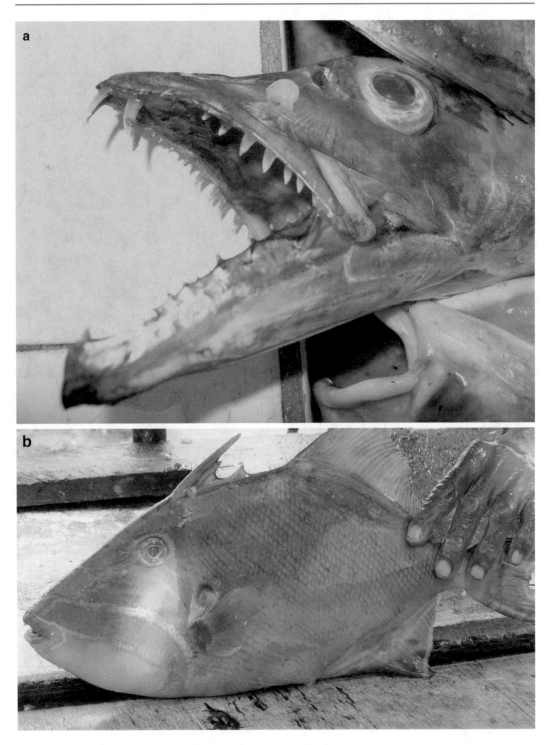

Fig. 3.122 (**a**) A swordfish (*Trichiuris lepturus*). (Photo: Vidal Haddad Jr.). (**b**) Triggerfish have a dorsal spike that can be "armed" by means of specific musculature and used to defend the fish. (Photo: Vidal Haddad Jr.)

Fig. 3.123   A grouper. (Photo: Fábio Lang da Silveira)

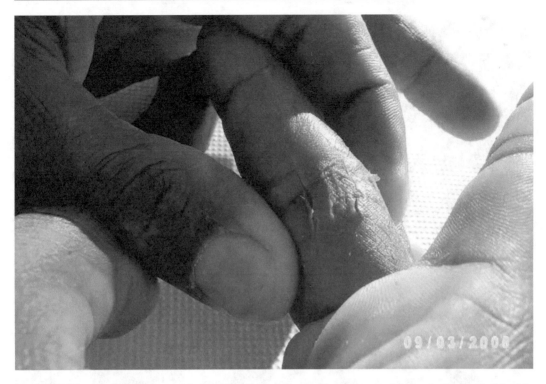

Fig. 3.124 These lacerated lesions on the hand of a diver were caused by a grouper when the man tried to feed the fish. (Photo: Fábio Lang da Silveira)

Fig. 3.125 This fish (*Xiphias gladius*) attacked a bather in shallow waters. (Photo: Lifeguard Troops of Matinhos, Matinhos, Paraná State, Brazil)

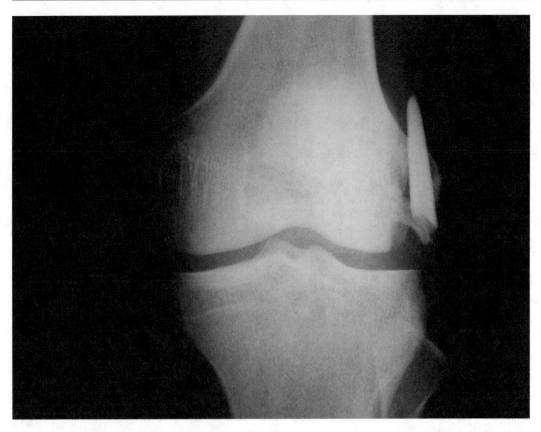

Fig. 3.126 Radiological examination of the knee of a victim, showing a fragment of the "beak" of a fish in the wound. (Photo: Hospital do Trabalhador of Curitiba City, Curitiba, Paraná State, Brazil)

Fig. 3.127 The pirambeba (*Serrasalmus maculatus*). This species of piranha is most commonly associated with bites suffered by people bathing in water bodies in Brazil. (Photo: Vidal Haddad Jr.)

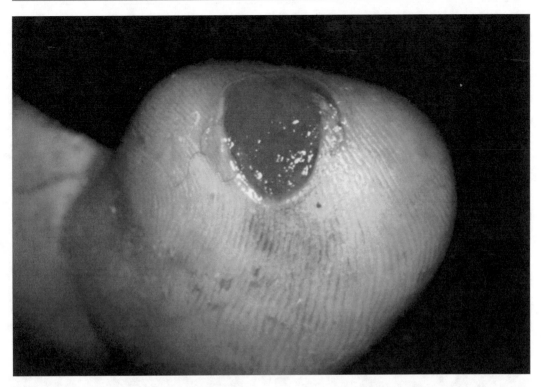

Fig. 3.128 A crateriform or "punch" lesion caused by a piranha bite on a bather. This lesion was the only bite and was presumably a warning to respect the territory and the eggs of the fish. (Photo: Vidal Haddad Jr.)

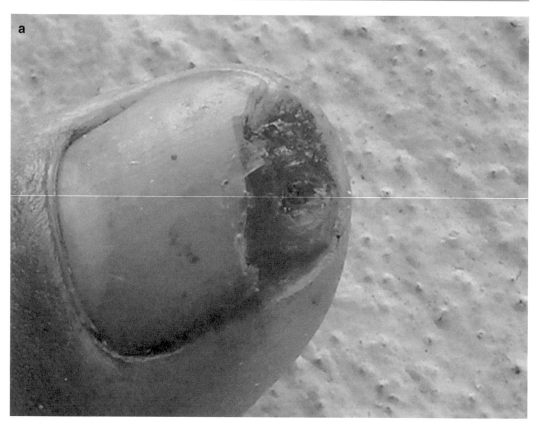

Fig. 3.129 (**a**) A "punch" lesion on the first finger of a bather, who was bitten in the water of a lake. (Photo: Vidal Haddad Jr.). (**b**) A bite with a crateriform appearance on a bather who was swimming in a dammed river tributary. During the same period when this injury occurred, several other attacks also occurred in bathers in the same location. (Photo: Isleide Saraiva Rocha). (**c**) *Left:* A recent bite suffered by a bather. *Right:* A scar showing the shape of the fish's dental arch. (Photos: Vidal Haddad Jr.)

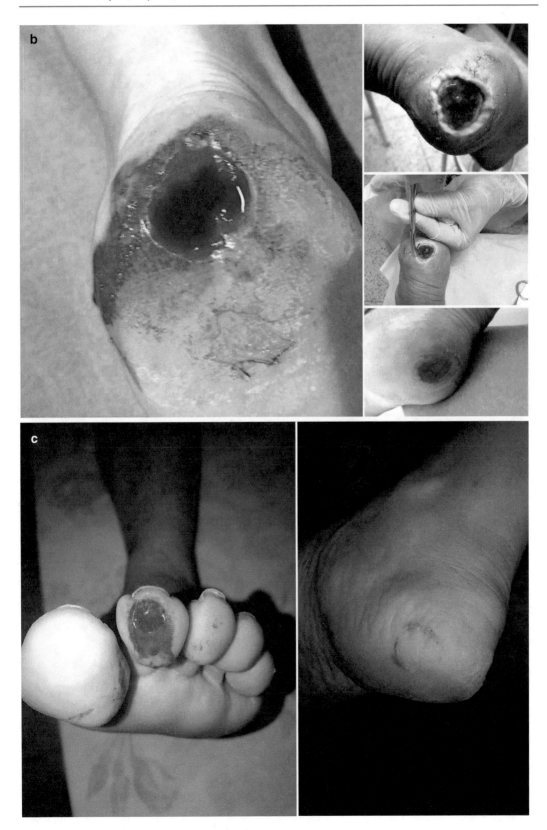

Fig. 3.129 (continued)

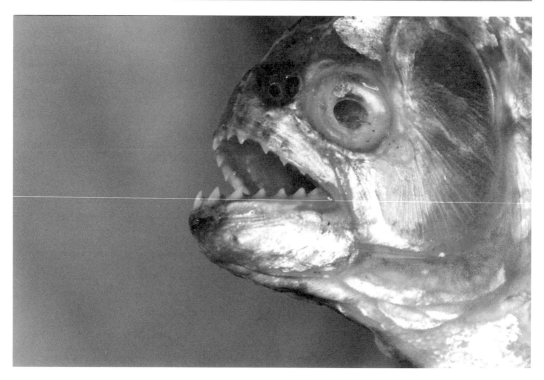

Fig. 3.130 Dentition of a specimen of *Serrasalmus maculatus*. (Photo: Vidal Haddad Jr.)

There are various genera of piranhas in South America. The genus *Serrasalmus* is found in all regions of Brazil, but the most important genus is *Pygocentrus*, which is present especially in the Amazon region and in the Brazilian Pantanal. The rare attacks caused by shoals appear to be associated with this genus.

The most common species in this genus is *Pygocentrus nattereri* (the red piranha, or cashew piranha). Its shoals tend to remain in flooded forests and lakes at the end of a drought period, causing danger to bathers in those waters (Fig. 3.131) [40].

In truth, almost all species of fish have the potential to cause injuries on the hands and feet of fishermen, especially amateur fishermen (Fig. 3.132). Such injuries are more likely to be caused by "sport" fish, catfish, large fish with sharp teeth (such as the golden dorado [*Salminus brasiliensis*]), and all other fish with similar characteristics [4, 5].

The piraíba (*Brachyplatystoma filamentosum*), pirarara (*Phractocephalus hemiliopterus*), and surubim (*Pseudoplatystoma corruscans*, *Pseudoplatystoma fasciatum*, and *Pseudoplatystoma reticulatum*) are large catfish capable of causing wounds in human beings. The surubim, pintado, and cachara are actually considered venomous fish because toxins have been found in their dorsal and pectoral stingers (Figs. 3.133 and 3.134).

The piraíba is the largest freshwater fish in South America and can reach 3 m in length and over 200 kg in weight (Fig. 3.135) [4, 5]. These fish have been accused of devouring human beings in the Amazon region, especially children swimming in deep waters. A testimony obtained by a physician (Dr. João Baptista de Paula Neto) in a indigenous village in Tocantins State, Brazil, described the discovery of the corpse of a 3-year-old child in a captured piraíba. Another testimony from a priest in Mato Grosso State, Brazil, reported an examination of the body of an indigenous 8-year-old child, who was described by a witness as having disappeared after being grabbed by a "big fish"; his body was found 3 days after the accident. His lower half was uninjured, but there were marks on his upper half "as if the skin had been sanded. His eyes were bulging, perhaps as a result of the pressure of having been sucked." The jaws of the great catfish do have a sandpaper-like aspect, and the injuries they could cause would be similar to those described by the priest. Although these reports were detailed, there was no photographic documentation of these accidents. There have been descriptions of human beings being devoured by big catfish in Asian rivers and even in Eastern Europe, with documented cases of children being swallowed by the wels catfish, or sheatfish (*Silurus glanis*), a large European catfish capable of measuring 5 m in length and weighing 350 kg [4].

Although the victims may have drowned before being eaten, the large diameter of the mouth of these fish makes it possible for them to occasionally capture small human beings. However, it is very difficult to verify such cases; in fact, there have been no documented cases of attacks on humans by these fish.

Of all venomous fish, the Siluriformes are the ones that most frequently cause injuries worldwide. In Brazil, they are certainly the fish most commonly associated with injuries in bathers, fishermen, and other professionals working in fishing communities in marine and freshwater environments. The venom of these fish causes severe pain, and secondary infections are very common; however, immediate measures such as use of hot water have good effects in controlling the pain (Table 3.1).

The true candirus also belong to the Siluriformes order (catfish). The Trichomycteridae family includes the subfamilies Stegophilinae and Vandelliinae; the former is a carnivorous fish that can bite humans, and the latter is hematophagous (Figs. 3.136, 3.137 and 3.138). These fish have a cylindrical and elongated body, with thin and sharp teeth [4, 5].

The candiru genera that have clinical significance are *Branchiocca*, *Paravandellia*, *Paracanthopoma*, *Vandellia*, and *Plechtrochilus* (formerly known as *Urinophilus*), which are all part of the Vandelliinae subfamily. They are small, slender, bloodsucking fish, which parasitize the

Fig. 3.131 The red piranha, or cashew piranha (*Pygocentrus nattereri*), is aggressive and can form shoals. In very rare situations, it may attack humans and other animals that agitate and/or bleed in the water. (Photo: Vidal Haddad Jr.)

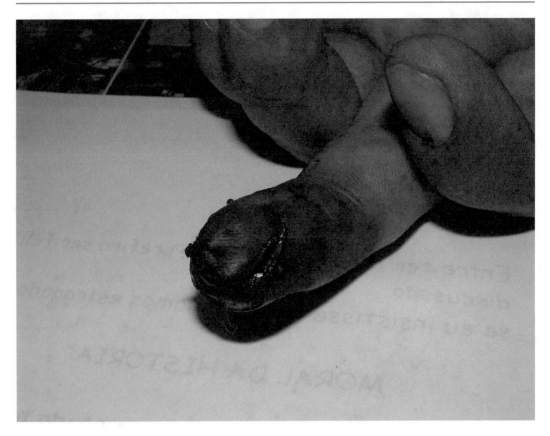

Fig. 3.132 A bite on the finger of a fisherman. Note the laceration and the secondary infection. (Photo: Vidal Haddad Jr.)

Fig. 3.133 A surubim can cause severe lesions and envenomation with its stingers. This specimen is *Pseudoplatystoma reticulatum* (also known as the cachara). Note the stinger in the detail. (Photo: Vidal Haddad Jr.)

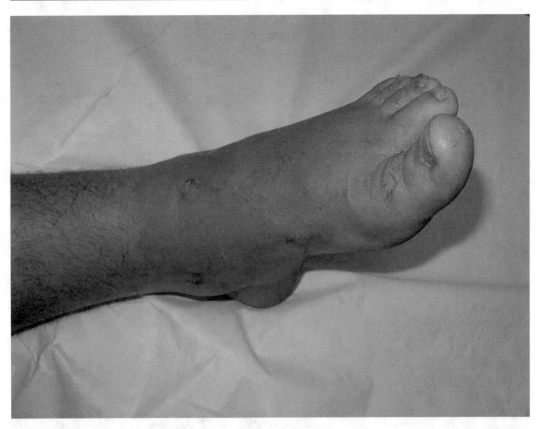

Fig. 3.134 This patient was wounded by a Surubim cachara and suffered intense pain and inflammation at the site of the sting. Two days after the injury, the site become more inflamed, with a purulent secretion, and the patient developed a fever (erysipelas)

Fig. 3.135 Piraíba (*Brachyplatystoma filamentosum*) in a fish market in the Amazonian region. (Photo: Vidal Haddad Jr.)

Table 3.1 Algorithm for identification and treatment of injuries caused by aquatic animals

	Puncture wound			Skin eruption		Lacerated wound	
	Presence of a stinger[a]	Localized presence of spines[b]	(Only rarely) presence of spines[a]	Urticariform plaque, edema, erythema, vesicles, necrosis[a]	Eczema-like lesion[b]	Cyanotic or pale wound edges, presence of stinger fragments[a]	Lacerations with pain proportionate to the wound[b]
Animal	Marine or freshwater catfish, stingray	Sea urchin	Scorpionfish, toadfish	Jellyfish, Portuguese man-of-war, coral, anemone	Marine or freshwater sponge, marine worm, sea cucumber	Marine or freshwater stingray or catfish (occasionally causing a puncture wound)	Shark, barracuda, moray eel, piranha, or other traumatogenic fish
Recommended treatment	Treatment 1	Treatment 1	Treatment 1	Treatment 2	Treatment 2	Treatment 1	Treatment 3

Adapted from Haddad [4]

Treatment 1: Immerse the wound in hot water (at a temperature of about 50 °C; test it with your hand) for 30–90 minutes. Remove any spines, stingers, or glandular epithelium fragments after infiltrating the wound with a local anesthetic agent. If symptoms persist at a late stage, perform a radiological examination and assess whether tetanus prophylaxis is required. In all cases of a lacerated wound, also assess whether an antibiotic should be administered (cefalexin [cephalexin] 2.0 g/day for 10 days or amoxicillin [amoxycillin] clavulanate 1.5 g/day for 10 days)

Treatment 2: Wash the site of the injury and make compresses with cold seawater; *do not use freshwater!* Then dress the wound with vinegar compresses. If analgesia is required, administer one ampule of dipyrone (also known as metamizole sodium) intramuscularly

Treatment 3: Wash the wound thoroughly and perform a surgical exploration of it. Assess whether tetanus prophylaxis is required. In all cases of a lacerated wound, also assess whether an antibiotic should be administered (cefalexin [cephalexin] 2.0 g/day for 10 days or amoxicillin [amoxycillin] clavulanate 1.5 g/day for 10 days)

[a] Accompanied by intense pain

[b] Accompanied by moderate pain

Fig. 3.136 Candirus of the Vandelliinae subfamily. These fish are hematophagous and can penetrate natural orifices in humans. (Photo: Vidal Haddad Jr.)

Fig. 3.137 The Stegophilinae subfamily consists of carnivorous fish that attack corpses and, rarely, live animals. (Photo: Vidal Haddad Jr.)

Fig. 3.138 *Top:* The Cetopsidae family includes the whale candirus, which are not true candirus. They are voracious fish, which attack drowned animals and can bite live humans. *Bottom:* Candirus of the Stegophilinae subfamily and the (smaller) Vandelliinae subfamily. (Photo: Vidal Haddad Jr.)

gills of larger fish, especially big catfish of the Pimelodidae family. Because of their sensitivity to the smell of ammonia or blood in the water, candirus may be attracted to urine or blood, and they may invade the urethra or other human natural orifices. The fish then extends its intraopercular "claws" (the odontoids) and cannot retract them. It then dies, causing obstruction and uremia or severe bleeding, which can cause the death of the victim.

Recently, there was a report of an attack on a human by a vandelliine candiru, which inflicted a wound on the body of its victim. This candiru was identified as a scientifically undescribed genus and species, and it was referred to as the "human-biting candiru." This fish bites its victim generally on the body, fastens itself there with its specialized teeth (and perhaps interopercular spines as well), and draws blood. Because of the biting force that this fish's powerful head muscles exert (Fig. 3.139), it is difficult to remove from the victim [41].

These injuries always require surgery, and candirus are much feared by riverside communities and swimmers in Amazonia. The Stegophilinae subfamily consists of carnivorous fish, which mostly attack corpses and only rarely attack live animals.

Whale candirus are larger fish (up to 30 cm long) and prey on shoals of other fish such as piranhas. They are not true candirus and are part of the Cetopsidae family. Whale candirus, in large numbers, penetrate the bodies of drowned humans via their natural orifices and feed on them from the inside out, consuming the viscera and then the muscles. Corpses attacked by whale candirus manifest numerous holes on their surface, through which the fish exit the body (Figs. 3.140 and 3.141) [4].

Piracatinga catfish of the species *Callophysus macropterus* (part of the Pimelodidae family) are carnivores and attack the corpses of animals and humans in rivers in the Amazon. These fish are extremely voracious and are commercialized in large numbers. They are caught with bait consisting of the meat of freshwater dolphins and alligators, which puts the populations of these animals at risk. For this reason, piracatinga fishing is now prohibited.

Piracatinga have been described as being of forensic value; in one case, the finding of human remains in the piracatinga digestive tract made it possible to identify a missing person after a probable drowning (Fig. 3.142) [42].

Traíras/trairões (*Hoplias* sp., also known as the South American snook), and pikes are aggressive fish with sharp teeth. They are distributed widely in freshwater bodies and account for a large number of bites and lacerated wounds in fishermen (Fig. 3.143). The peacock bass (*Cichla* sp.) is another fish that can inflict puncture wounds or lacerations with the rays on their dorsal fins, and these wounds frequently become infected (Figs. 3.148 and 3.149).

Some attacks on humans in freshwater environments—even those thousands of kilometers upriver from the sea—are caused by the bull shark (*Carcharhinus leucas*) [4, 5].

The electric eel (*Electrophorus electricus*) is capable of delivering electric shocks of up to 600 V to animals and humans in the water (Fig. 3.150). Although this voltage is hardly capable of killing a human being, the victim can drown as a result of muscle contractions and stiffness caused by the electric shock. The author has previously observed a human corpse in the Amazonian region; witnesses who were with the victim in the river described seeing the electric eel and observed the shock it delivered. However, the victim's necropsy showed only water in the lungs, with no signs of an electric shock. This fish is feared in the areas it inhabits in the Amazon.

Reptiles (Turtles, Alligators, Crocodiles, and Snakes)

Aquatic turtles can be found in fluvial and marine environments. Some freshwater turtles are known to be aggressive, such as the North American snapping turtle (*Chelydra serpentina*), a large animal (measuring up to 50 cm in length) that can cause serious traumatic injuries. Its cousin the alligator turtle (*Macrochelys temminckii*) is even bigger but less aggressive. There are several sea turtles in the Cheloniidae family; some (leatherback turtles) can reach a weight of 700 kg and measure more than 2 m long. These large animals

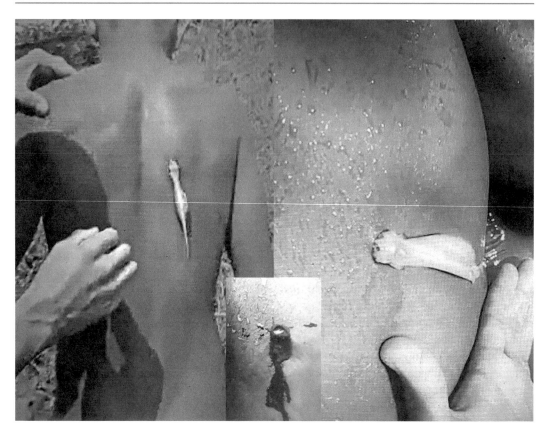

Fig. 3.139 This candiru attached itself to its victim to suck blood. When it was removed, its mouth parts left a deep oval wound. (Photo: Vidal Haddad Jr.)

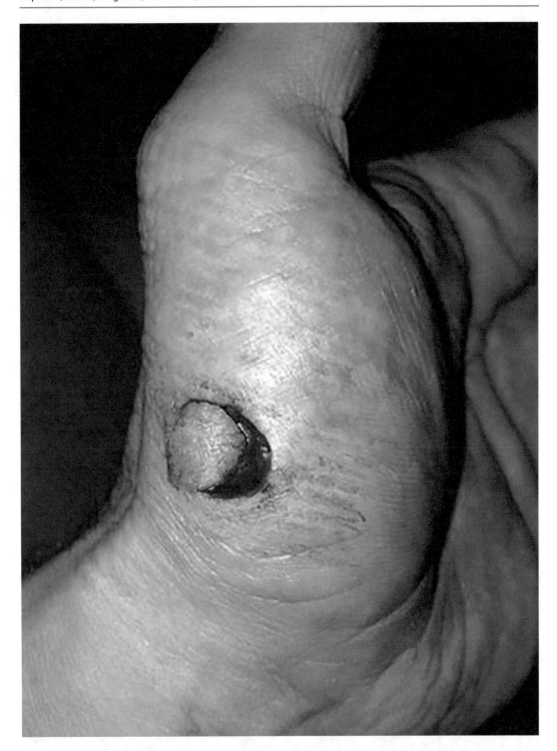

Fig. 3.140 The lesion caused by a candiru in a diver that was in a rescue mission in an Amazonian River

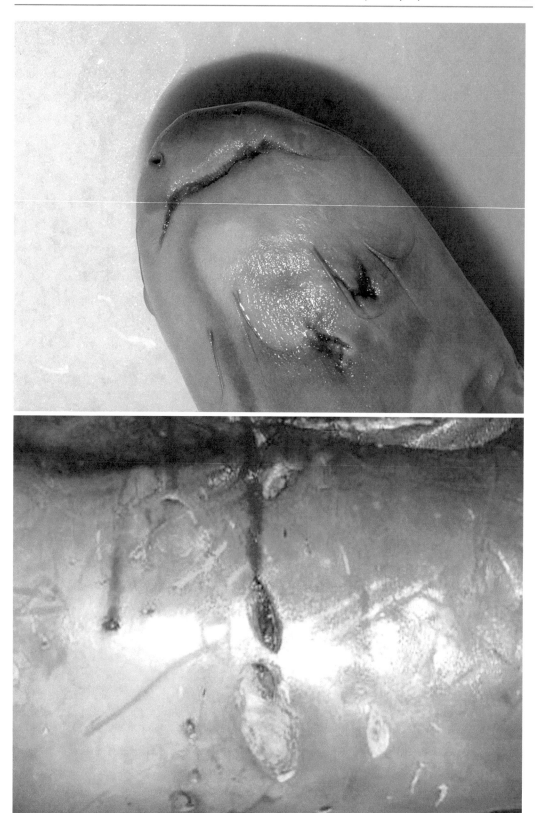

Fig. 3.141 "Punch" lesions, caused by Stegophilinae candirus, on the corpse of a drowning victim. Note the shape of the mouth of a candiru. (Photo: Domingos Garrone Neto)

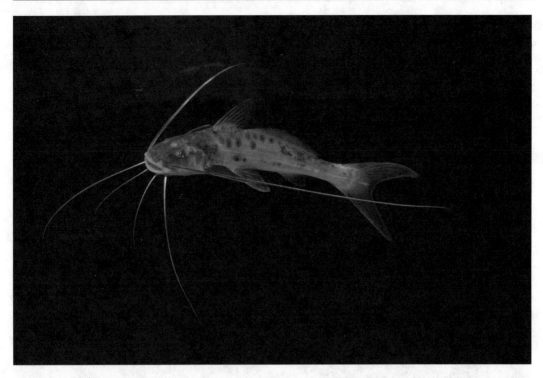

Fig. 3.142 The piracatinga is a carnivorous and voracious Amazonian catfish, which in shoals attacks corpses and has value in Legal Medicine, by analyzing the contents of their digestive tract. (Photo: E. Ferreira & J. Zuanon, INPA, Amazonas)

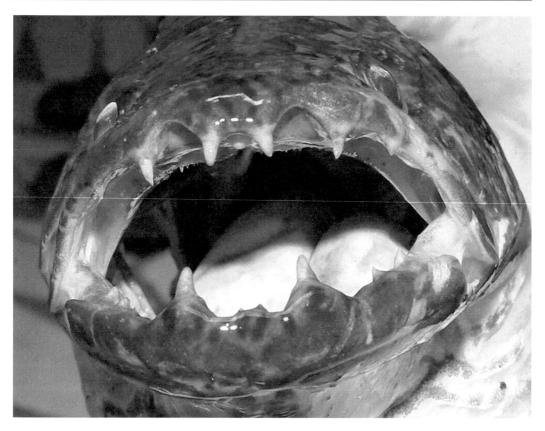

Fig. 3.143 Dentition of a traíra (*Hoplias malabaricus*). This fish is responsible for a large proportion of the fish bites that occur in South America. (Photo: Vidal Haddad Jr.)

Box 3.7 Piranhas

Two males aged 13 and 26 years and a woman aged 37 years were treated at an emergency room in Adolfo, São Paulo State, Brazil, after being bitten on the foot by piranhas while swimming in a pond, which was used by tourists and had been formed by damming a small local river (Figs. 3.144 and 3.145). This place was much frequented on weekends, and 2 months previously, piranha bites had started to happen there. Since then, more than 150 bathers had been bitten, some incurring significant bleeding and extensive lacerations. Each of these three patients had a single "piecemeal" wound, which was round to oval in shape, with intense bleeding that was controlled by localized compression. The site was extensively washed and treated with a pressure dressing and a topical antibiotic. The wound healed completely within about 20 days.

Comments: Human wounds caused by piranhas do not have the profile of an attack by a shoal, which is very rare in humans. Events in which multiple humans each receive a single bite are compatible with male piranhas defending their eggs laid in aquatic vegetation; this has been described as the main factor involved in piranha attacks. The wounds are deep, and treatment should be focused on controlling bleeding and possible secondary infections.

Fig. 3.144 Two piranhas of different genera: *Serrasalmus marginatus* and *Pygocentrus nattereri*. The latter can form aggressive shoals, but attacks on live humans are very rare. (Photo: Vidal Haddad Jr.)

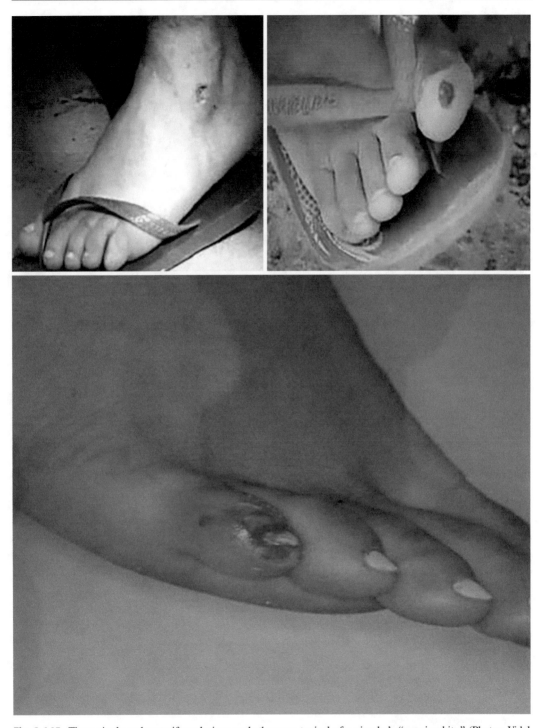

Fig. 3.145 These single and crateriform lesions on bathers are typical of a piranha's "warning bite." (Photos: Vidal Haddad Jr. and TV Tem)

Box 3.8 Candirus

In a case treated by Dr. Anoar Samad in Manaus (Amazonas State, Brazil), a 23-year-old man sought treatment at an emergency room, complaining that a fish had entered his urinary tract 1 day previously when he had urinated while standing in the Negro River; since then, he had been unable to urinate. On examination, the patient had severe inflammation of the glans penis, and his general condition was not good, with pallor, fever, severe pain, and bleeding from the penis. He underwent cystoscopy, and after 2 hours of surgery, a fish 12 cm long by 1.5 cm wide was removed. It was identified as a *Plectrochilus* (Siluriformes order, Trichomycteridae family) (Figs. 3.146 and 3.147).

Comments: These fish seem to be attracted by the odor of ammonia, which is similar to the smell of the gills of large fish, in which it is a parasite (hematophagous). This patient was very lucky to be close to a medical center and a specialist. Such accidents in distant parts of the Amazon may result in deaths.

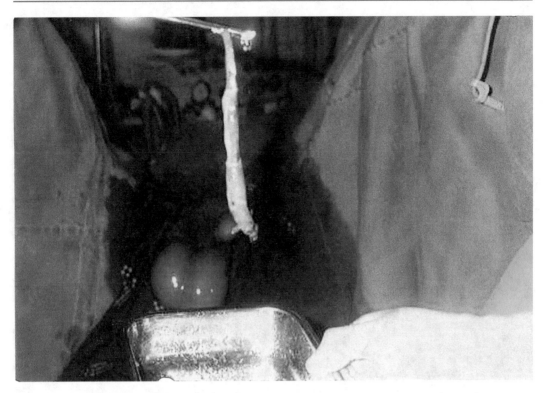

Fig. 3.146 This rare image shows extraction of a candiru from the urethra of the victim described in this clinical case. Note the edema and erythema of the glans. (Photo: Dr. Anoar Samad, Instituto de Urologia, Manaus, Amazonas State, Brazil)

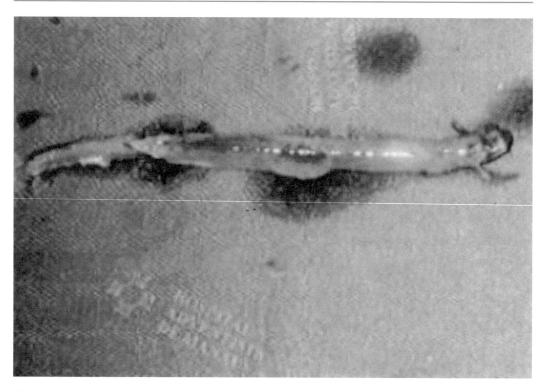

Fig. 3.147 The candiru specimen extracted from the penis of the patient was identified as being a *Plechtrochilus* (part of the Vandelliinae subfamily). (Photo: Dr. Anoar Samad, Instituto de Urologia, Manaus, Amazonas State, Brazil)

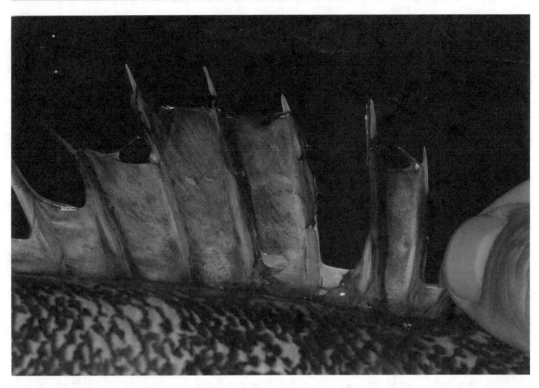

Fig. 3.148 The dorsal fin of the peacock bass (also known as the tucunaré) has sharp rays, which cause injuries in fishermen. (Photo: Vidal Haddad Jr.)

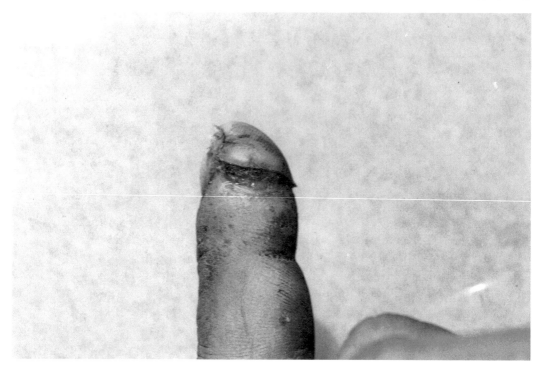

Fig. 3.149 A lacerated lesion on the finger of an amateur fisherman, caused by the rays on the dorsal fin of a peacock bass. (Photo: Vidal Haddad Jr.)

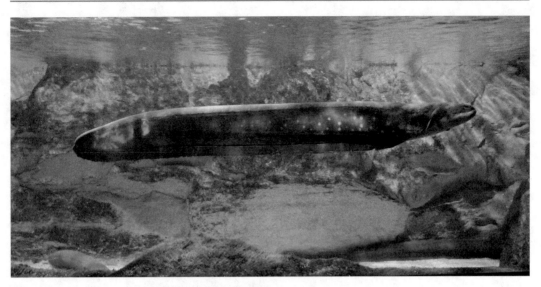

Fig. 3.150 An electric eel (*Electrophorus electricus*). (Photo: Vidal Haddad Jr.)

are peaceful and rarely injure humans, but fishermen and researchers are occasionally bitten by them (Fig. 3.151).

Some reptiles live in aquatic or semiaquatic locations. All crocodile and alligator species may cause human injuries with their teeth, tails, and speed of movement in the water. Most such accidents and attacks occur in the Amazon, the Nile River, and the Indo-Pacific region, and are caused by the species *Melanosuchus niger* (the jacaré-açu, or black alligator), *Caiman crocodilus* (the jacaretinga, or speckled caiman), *Alligator mississippiensis* (the American alligator), *Crocodylus niloticus* (the Nile crocodile), and *Crocodylus porosus* (the saltwater, estuarine, or Indo-Pacific crocodile) (Figs. 3.152, 3.153, 3.154, and 3.155) [4, 43, 44].

The Amazon jacaretinga and the black caiman are frequently associated with injuries in humans. Because of its man-eating reputation (which is not entirely accurate) and demand for its skin, the black caiman has been hunted toward extinction. Their place in nature, especially in the Amazon region, has therefore been taken over by the jacaretinga, which reaches up to 3 m in length. An adult black caiman can measures 4–6 m in length [43].

Injuries caused by crocodiles and alligators are very severe, including laceration and tearing of tissues, profuse bleeding, and serious infections caused by the animal's mouth flora (Figs. 3.156, 3.157, 3.158, and 3.159). Attacks by crocodilians show predatory tactics: they attack before anyone present detects their presence, illustrating a silent form of approach. With the victim trapped in its mouth, the animal submerges to drown it and keeps it under the water for hours, precipitating early decomposition to facilitate subsequent disarticulation of the victim [4, 43, 44].

Reported attacks have mainly involved fishermen (pulling up nets or diving for objects), bathers, or people working on riverbanks. The injuries caused by crocodiles and alligators cover the same spectrum as injuries caused by sharks, are also potentially fatal because of the complications described above, and require a similar therapeutic approach. Medical care for a victim of a crocodilian injury should be started as early as possible, with intensive washing of the wound, surgical debridement, and treatment of bleeding and possible fractures, since the jaws of the reptile close with tremendous force and can crush bones. Immediate institution of antibiotic therapy is very important to prevent serious infections [43, 44].

Various venomous and nonvenomous snakes live in aquatic and semiaquatic environments. Some of them are associated with human injuries and envenomations. Unlike their representations in legends and movies, the anacondas of the Boidae family are not aggressive snakes, but they do occasionally wound humans when provoked. There are four species of anaconda: *Eunectes murinus* (the green anaconda), *Eunectes notaeus* (the yellow anaconda of the Pantanal), *Eunectes deschauenseei* (which is restricted to Marajo Island in Pará State, Brazil) and *Eunectes beniensis* (the Bolivian anaconda) (Figs. 3.160, 3.161, 3.162, and 3.163). These snakes are large (up to about 7 m in length) and have the ability to suffocate their prey and then swallow them whole, but they rarely attack humans [4, 5].

In the author's experience, the majority of reported cases of attacks by anacondas are not factual. Images on the Internet of humans who were supposedly devoured by anacondas are, in fact, those of people whose deaths were caused by the reticulated python (*Python reticulatus*) in the Indo-Pacific region. Pythons are giant snakes of the Pythonidae family, which can measure up to 8 m in length and really are capable of devouring humans. However, there have been some cases of real anaconda attacks, such as one that occurred in early 2007, when an anaconda approximately 5 m in length attacked a child next to a stream without any provocation. The child was saved by his grandfather, and the snake possibly attacked for the purpose of predation. These attacks, however, are very rare, and—unlike pythons—no anaconda has ever been scientifically proved to have devoured a human being. The injuries caused by alligators and giant snakes are lacerations with deep wounds and occasionally loss of tissue (Fig. 3.164) [4, 5].

Some venomous snakes live in aquatic and semiaquatic environments. The South American

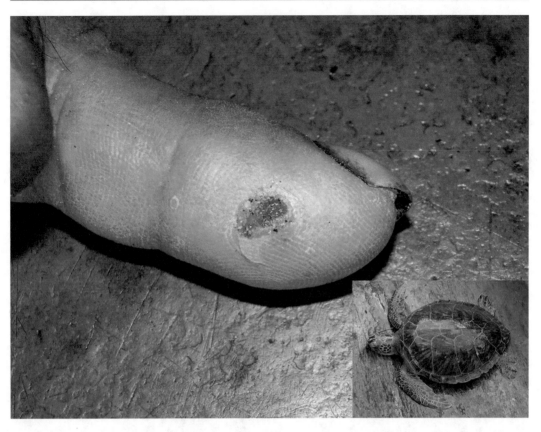

Fig. 3.151 Sea turtles may occasionally bite fishermen, as in this case. The victim was handling a green turtle in an attempt to free it from a fishing net. (Photos: Vidal Haddad Jr.).

Fig. 3.152 This mammoth black caiman (measuring 5.5 m in length) was captured in Brazil at a time when the country had no wildlife protection rules. (Photo: unknown photographer).

Fig. 3.153 *Top:* A black caiman (*Melanosuchus niger*). *Bottom:* A speckled caiman (*Caiman crocodilus*). (Photo: Vidal Haddad Jr.)

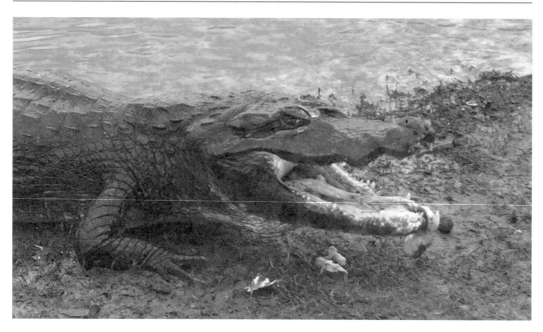

Fig. 3.154 The Pantanal jacaré (Caiman yacare) is a medium-sized caiman, which rarely attacks humans. (Photo: Vidal Haddad Jr.)

Fig. 3.155 The Nile crocodile (*Crocodylus niloticus*) is one of the most feared predators in the world. It is responsible for large numbers of attacks on humans and consequent fatalities. (Photo: Vidal Haddad Jr.)

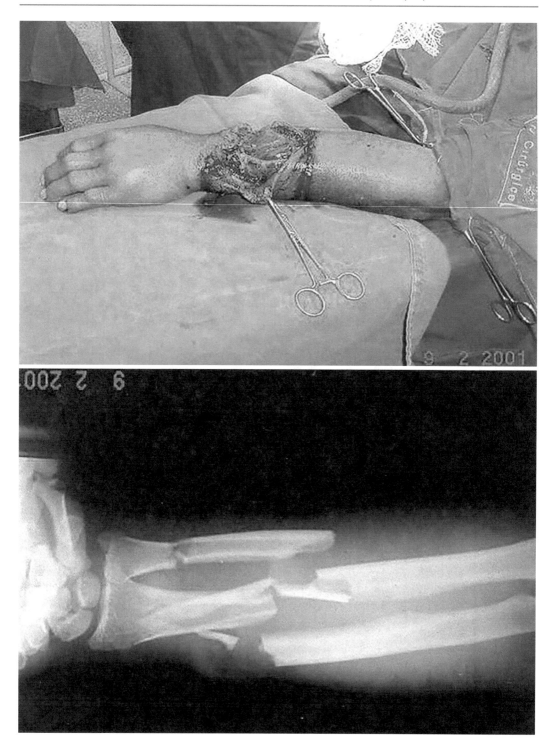

Fig. 3.156 A violent attack by a black caiman on a fisherman gathering a fishing net immersed in a river in the Amazon region shows the force of the reptile bite, which is capable of causing fatalities. (Photo: Dr. Nelson Henrique C. Oliveira, Hospital Universitário Getúlio Vargas, Manaus, Amazonas State, Brazil)

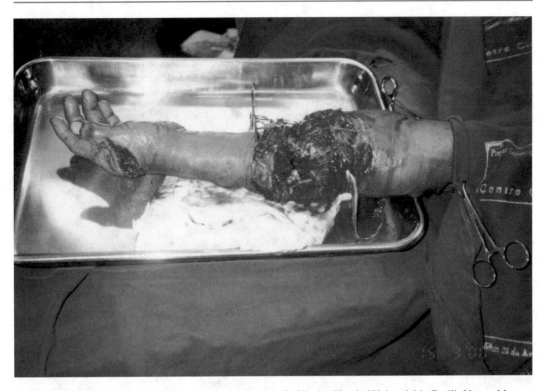

Fig. 3.157 The destructive bite of a black caiman caused extensive lacerations on this fisherman and also ruptured muscles and tendons. (Photo: Dr. Nelson Henrique C. Oliveira, Hospital Universitário Getúlio Vargas, Manaus, Amazonas State, Brazil)

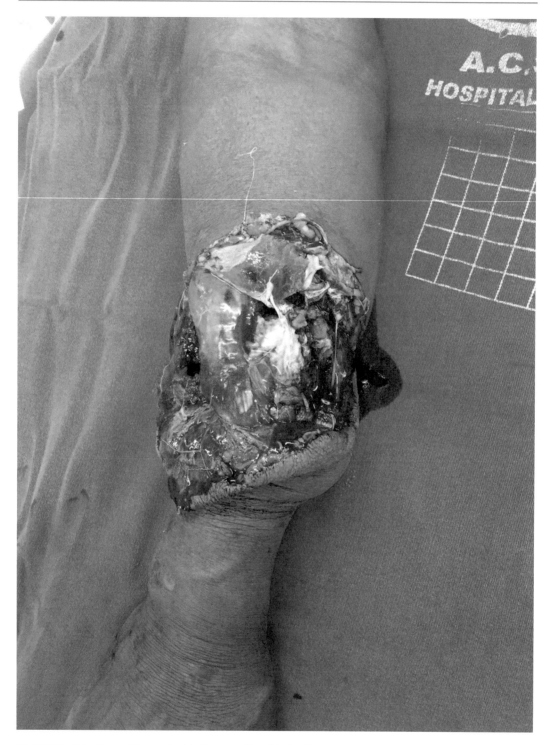

Fig. 3.158 Extensive and destructive injuries caused by a black caiman (*Melanosuchus niger*), which attacked a fisherman removing a fishing net from a river in the Amazon region. (Photo: Vidal Haddad Jr.)

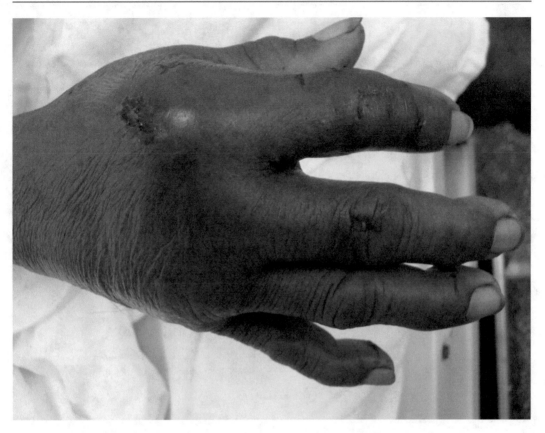

Fig. 3.159 The hand of this amateur fisherman was bitten by a Pantanal caiman when he was putting fish in a basket tied to a boat. This type of attack is really an accident, as the caiman did not intend to attack the human. (Photo: Dr. Manoel Francisco Campos Neto, Cáceres, Mato Grosso State, Brazil)

Fig. 3.160 A green anaconda (*Eunectes murinus*). This species of snake is the heaviest in the world and also one of the longest. (Photo: Vidal Haddad Jr.)

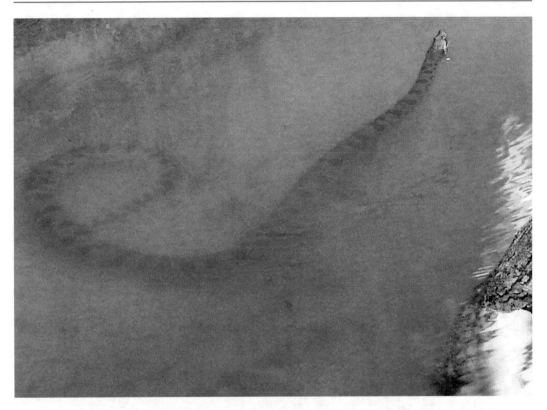

Fig. 3.161 AA green anaconda in the water. Anacondas hunt prey in the water (where they can move very fast) or on the banks of water bodies. (Photo: Vidal Haddad Jr.)

Fig. 3.162 The heads of the two largest constricting snake species in the New World. *Top:* A boa constrictor (*Boa constrictor*), which can reach up to 4 m in length. *Bottom:* A green anaconda (*Eunectes murinus*), which can be 6–7 m long

Fig. 3.163 The yellow anaconda (*Eunectes notaeus*) is a smaller species of anaconda found in the Pantanal region in Brazil and in the Chaco region in Paraguay and Bolivia. (Photo: Vidal Haddad Jr.)

Fig. 3.164 Anacondas can eat medium-sized animals, such as this calf. There have been no proven reports of humans being devoured by anacondas. (Photo: unknown photographer)

jararacucu (*Bothrops jararacussu*), a member of the Viperidae family, is a large venomous snake, which can reach 2 m in length and inject up to 1 g of proteolytic and coagulant venom (Fig. 3.165). These snakes live near water bodies, where their bites occur. There has been one report of a human death; it occurred about 10 minutes after a bite from a jararacucu, which caused an arterial puncture.

Other species of *Bothrops* (Viperidae) also live near aquatic environments. An example is the jararaca (*Bothrops jararaca*), which is found in South and Central America. It is responsible for the majority of envenomations in the region and is much feared (Fig. 3.166). The venom of these snakes, which can be fatal to humans, causes intense localized inflammation, skin necrosis, and alterations in blood coagulation (Figs. 3.167 and 3.168).

The coral snake (*Micrurus surinamensis*) is a semiaquatic snake found in northern South America. Like all members of the Elapidae family, it has a very dangerous bite because its venom is neurotoxic.

The recommended treatment for venomous snakebites is administration of antivenom [4].

Sea snakes are common animals in some parts of the Indo-Pacific region. There are doubts about the zoological classification of these snakes. Some authors consider these highly venomous reptiles part of the Elapidae family, whereas others consider them part of the Hydrophiidae family. They are separated into two groups, which are based on the terrestrial snakes from which these subfamilies originated: the elapids of the Australian region are related to the Hydrophiinae subfamily (also called true sea snakes – Fig. 3.169) and the cobras of Asia are linked to the sea kraits of the Laticaudinae subfamily (Fig. 3.170). Although their venom is highly toxic, envenomations are rare, as the fangs of sea snakes are very short and these animals are not aggressive. There have been rare reports of injuries in fishermen. True sea snakes are adapted to the aquatic life (with a flattened tail, valvular nostrils, and elongated lungs), but sea kraits live partially on land, as they are not totally adapted to it. All species of sea snakes are viviparous, but sea kraits are oviparous, laying their eggs on land [45, 46].

There are nearly 60 species of sea snake, which feed on fish and fish eggs. They are not large reptiles; most specimens measure about 1 m in length. The yellow-bellied sea snake (*Hydrophis platurus*, formerly known as *Pelamis platurus*) can be found throughout the entire Indo-Pacific region, spreading to the western coasts of the Americas (from the Galapagos Islands to the southern USA) (Figs. 3.171, 3.172, and 3.173) [46–51].

Envenomation caused by sea snakes is severe. The bite is painless, and its main effects are intense rhabdomyolysis associated with myotoxicity and muscular paralysis caused by neurotoxicity. Only minor inflammation may be visible at the site of the bite. Systemic manifestations begin 30–120 minutes after the bite occurs. Initially, it is possible to observe cephalgia, cold sweats, nausea, and vomiting. In the later stages, there is malaise, muscular pain and weakness, and trismus. The envenomation can have very severe effects, such as progressive flaccid paralysis (marked by initial ptosis) and paralysis of the musculature, which can cause death if it affects the respiratory muscles. As a result of the rhabdomyolysis, myoglobulin is present in the urine, which shows red/brown/black coloration, and acute renal failure, hyperkalemia, and cardiac arrest can be precipitated.

The treatment of a sea snake bite includes initial pressure at the site of envenomation. Since the venom does not cause localized necrosis or inflammation, use of pressure dressings can slow the spread of the toxin and delay the onset of systemic manifestations. Polyvalent sea snake antivenom is available through CSL and should be administered as soon as possible to any patient with signs of envenomation. The effects of Australian tiger snake venom are very similar to those of sea snakes; thus, use of tiger snake antivenom is an option in the absence of sea snake antivenom [52]. The recommended doses of antivenom are one ampule (1000 U) of antivenom for mild/moderate envenomation and 3–10 ampules of antivenom for severe envenomation [50, 52].

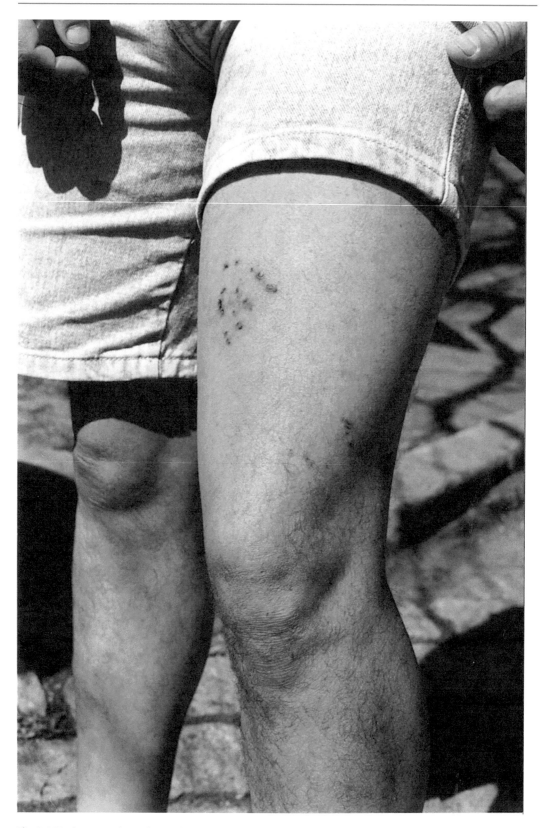

Fig. 3.165 A rare attack on a human by an anaconda. This amateur fisherman probably approached the hunting area or the nest of the anaconda, which reacted by attacking him. (Photo: Dr. João Luiz Costa Cardoso, Instituto Butantan, São Paulo, São Paulo State, Brazil)

Fig. 3.166 One of the most venomous Viperidae in South America is the jararacuçu (*Bothrops jararacussu*). Its venom causes intense localized inflammation with necrosis and alterations in blood coagulation. (Photo: Vidal Haddad Jr.)

Fig. 3.167 The jararaca (*Bothrops jararaca*). This genus of Viperidae is responsible for about 90% of the snakebites in South America. (Photo: Vidal Haddad Jr.)

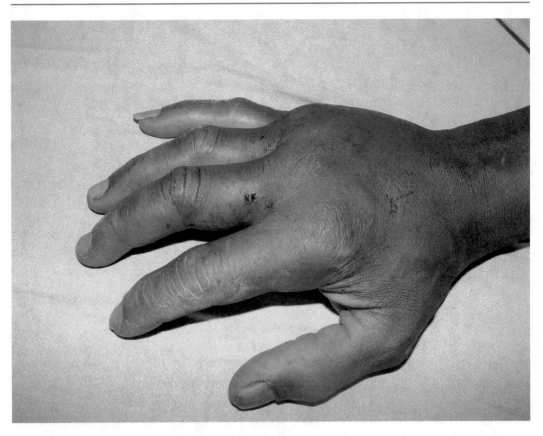

Fig. 3.168 Edema, erythema, and bite marks on the hand of a patient bitten by a jararaca (*Bothrops jararaca*). These manifestations are typical in the initial phases of envenomation. (Photo: Vidal Haddad Jr.)

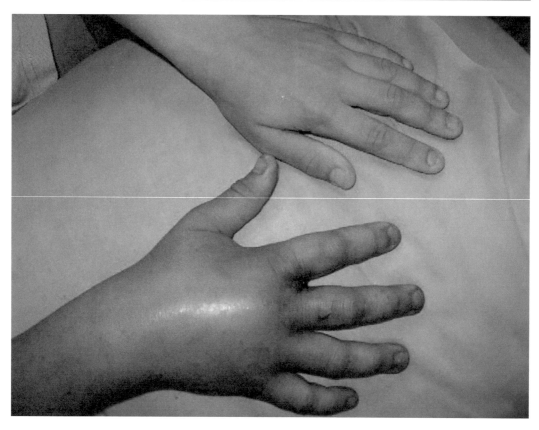

Fig. 3.169 Intense inflammation is part of the action of the venom of some Viperidae. This victim was bitten by a *Bothrops jararaca*. (Photo: Vidal Haddad Jr.)

Fig. 3.170 A Hydrophiidae snake (probably of the genus *Enhydrina*). This dangerous snake rarely causes envenomations in humans. (Photo: Vidal Haddad Jr.)

Fig. 3.171 Sea kraits (*Laticauda* sp.) are members of the Hydrophiinae family and are considered to have originated from the cobras of Asia. (Photo: Vidal Haddad Jr.)

Fig. 3.172 *Pelamis platurus* is the most disseminated species of sea snake, being founded in the Pacific Coast of Americas. (Photo: Dr Alejandro Solórzano, Costa Rica)

Fig. 3.173 The specimens of *Pelamis platurus*) show great variation of colors. (Photo: Dr Alejandro Solórzano, Costa Rica)

Fig. 3.174 The yellow-bellied sea snake is a beautiful but dangerous snake. In some parts of the world, large numbers of them can be found on the sands of beaches after a storm. (Photo: Dr. Alejandro Solórzano, Costa Rica)

Amphibians

Amphibians are fragile animals. This would already have resulted in their extinction by predators, were it not for the fact that many of them accumulate or produce toxins, which they use for defense.

Toads and tree frogs are the main users of these resources, and both live close to water bodies or in humid environments. Toads can be found in urban and forest environments, and tree frogs inhabit compact forests.

Toads have bufotoxins (as well as other toxins), which act through pharmacological mechanisms very similar to that of digitoxin, a glycoside that stimulates the heart rhythm, with the potential to cause deaths in humans and animals. Serious envenomations in humans are rare and are linked to use of the venom as a stimulant drug, which is taken by smoking dried toad skin. Poisonings in domestic animals are not uncommon, and they happen when toads are attacked and bitten in the parotid glands, which are located on the sides of the head (Fig. 3.174). The treatment is the same as that used for digitalis poisoning, which is not uncommon in patients with heart failure.

Poisonous tree frogs, or poison dart frogs, are the most beautiful representatives of poisonous animals in the world (Fig. 3.175). They accumulate alkaloids from ants, centipedes, and mites, and they synthesize toxins with neurotoxic effects, paralyzing smooth and striated muscles, and not infrequently causing deaths. Some Amazonian tribes apply cutaneous secretions from tree frogs to the blowpipes they use as hunting weapons. These small amphibians have very showy colors as part of a biological mechanism called aposematism, which "warns" predators of the danger posed by their toxins. The beauty of these animals is stunning, but the risk involved in handling them is great.

Kambô is a ritual that is part of the culture of some Amazonian tribes. They use secretions from the skin on the back of the tree frog *Phyllomedusa bicolor*, which are considered medicinal (Fig. 3.176). Green tree frogs feed on insects in the trees, and their tadpoles develop in temporary water bodies that collect in hollows in the trees.

Two indigenous Amazonian ethnic groups—the Katukinas and Kaxinawás—extract cutaneous secretions from the frogs and apply them to obtain analgesic and antibiotic effects. According to their culture, use of these extracts also helps to prevent disease. Phyllomedusin, phyllokinin, caerulein, and sauvagine are polypeptides that cause intense effects on smooth muscle vessels, with nausea, vomiting, arterial hypotension, flushing, palpitations, nausea, vomiting, bile secretion, and angioedema. The actions of these compounds are similar to those of bradykinin. The dose depends on the shaman and the patient. The healers themselves warn others about misuse of kambô, which, in inexperienced hands, can cause serious problems. Given the potent actions of these substances, as have been demonstrated, it seems clear that use of them by outsiders is very dangerous [53].

Mammals

There are rare aquatic mammals that use venom as a defense mechanism. As a curiosity, the semi-aquatic platypus—a mammal native to Australia—lays eggs and has a beak and feet similar to those of a duck. Moreover, the male has venomous spurs on its hind legs (Fig. 3.177). Envenomation by a platypus is not serious but does cause the victim pain for a few hours (Fig. 3.178) [4].

Marine mammals can, on rare occasions, cause traumatic injuries in humans (especially seals, sea lions, dolphins, and orcas). These attacks are often unprovoked, i.e., they are caused by handling of the animal or other situations that cause them stress. Thus, these accidents happen with orcas in aquarium exhibitions, such as the episodes that occurred with an orca named Tilikum. Between 1991 and 2010, he caused three fatal incidents at Sea World in Orlando (Florida, USA). Traumatic injuries can also be caused by seals and sea lions, and they are usually precipitated by the actions of tourists trying to touch or get close to the animals, although

Fig. 3.175 A cane toad (*Rhinella icterica*). This species, which is related to the genus *Bufo*, has large parotid glands on the sides of its head and can cause severe poisoning. (Photo: Vidal Haddad Jr.)

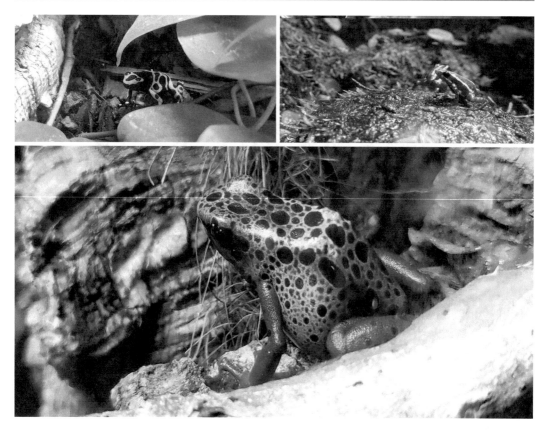

Fig. 3.176 Tree frogs of the family Dendrobatidae, with emphasis on the blue tree frog (*Dendrobates tinctorius*). (Photos: Vidal Haddad Jr.)

Fig. 3.177 Extraction of dorsal secretions from the *Phyllomedusa bicolor* tree frog are used to perform the kambô ritual. (Photo: César M. C. Pedigone)

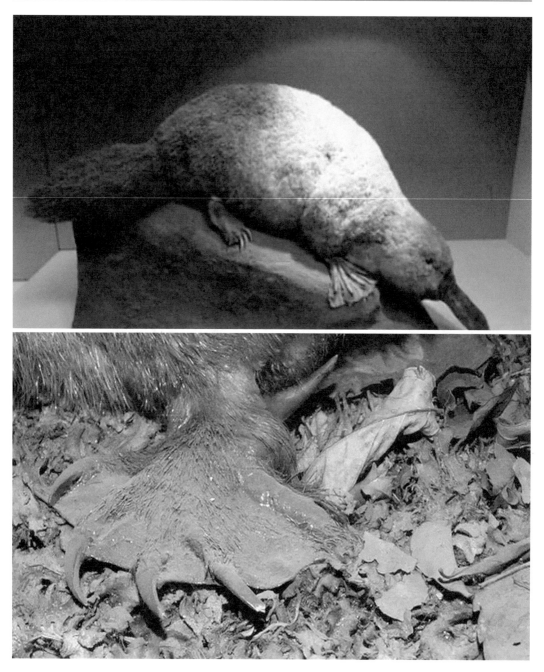

Fig. 3.178 The platypus is a peculiar mammal; the males have a venomous spur on their hind legs. Their stings can cause intense pain in victims, but envenomation is very rare. (Photo: Vidal Haddad Jr.)

there have been reports of attacks on bathers. Likewise, serious accidents can be caused by dolphins when bathers approach, touch, or capture animals close to beaches, as occurred in 1994 on the southeastern coast of Brazil, where a dolphin nicknamed Tião wounded eight people over a 15-day period while it was at a beach, being teased and touched by bathers. One person died from internal bleeding after a blow to the chest from Tião's tail [54].

Injuries caused by these mammals can be severe because of the size of the animals, and they should be treated with caution because the risks of bone fractures and injuries to internal organs are high and bleeding can cause the victim's death.

Treatment of Trauma and Envenomations by Fish

Treatments for injuries caused by fish are detailed in this section. The complications of these injuries can vary. Spines and stings may carry agents of secondary infection, including tetanus and sporotrichosis [55–57]. Fragments of spines and stingers can be retained in the wound, causing chronic inflammation and foreign body reactions [55–57].

Treatment should be administered as early as possible. The initial measures do not require hospital care [58, 59]. The pain of a wound caused by a venomous fish can be reduced by immersing the wound in hot water (with a temperature of about 50 °C) for 30–90 minutes. This follows the principle that all fish venom is thermolabile; however, the pain returns after the wound is removed from the hot water. This fact reinforces the hypothesis that whereas the venom promotes intense vasoconstriction and ischemia/necrosis/pain, the hot water induces vasodilatation and counterbalances the ischemia and the pain [4].

Exploration It is necessary to remove stinger fragments or epithelium fragments from the wound, under medical supervision with local anesthesia (to control the pain). Intramuscular administration of an ampule of dipyrone (also known as metamizole sodium) may be necessary to control the pain.

Other Measures Prevention of tetanus infection and radiological investigation (to look for stinger fragments) may be necessary. If there is severe systemic involvement (e.g., the presence of shock, cardiac arrhythmia, or respiratory failure), the victim must be transferred urgently to an intensive care unit.

The algorithm shown in Table 3.1 facilitates identification by hospital staff of most such accidents that occur around the world. After recognition of the causative agent, there are steps that can be applied at the time of the injury by any individual, and this knowledge is especially useful for lifesaving teams and paramedics. Thus, envenomation of any severity by a cnidarian can be treated immediately with a marine ice water bath and vinegar dressings. Other measures listed in Table 3.1 must be used in severe cases (such as possible use of mechanical ventilation, administration of verapamil for arrhythmias, administration of dipyrone application for pain control, etc.).

It is not always possible to apply therapeutic measures to wounds from envenomations caused by fish, particularly in remote locations; thus, it is essential to use hot water, as specified in Table 3.1 [58–63]. Other measures (such as extraction of sting or fragments, washing of wounds, etc.) should be performed in a hospital setting by experienced health professionals. In some cases, chronic ulcers may develop, especially on the lower limbs, and these may require skin grafts.

Finally, the number of species of aquatic animals that can cause injuries in humans is very large, and we have too little knowledge about them. More and more, this type of occurrence is presenting to hospitals and emergency medical centers, forcing health professionals and even ordinary people to have basic knowledge on this subject, which is rich in species, rich in beauty, and rich in dangers.

References

1. Haddad V Jr. Avaliação epidemiológica, clínica e terapêutica de acidentes provocados por animais peçonhentos marinhos na região sudeste do Brasil [thesis]. São Paulo: Escola Paulista de Medicina—UNIFESP; 1999.
2. Haddad V Jr. Atlas de animais aquáticos perigosos do Brasil: guia médico de diagnóstico e tratamento de acidentes. Editora Roca: São Paulo; 2000.
3. Haddad V Jr. Animais aquáticos de importância médica. Rev Soc Bras Med Trop. 2003;36(5): 591–7.
4. Haddad V Jr. Animais aquáticos potencialmente perigosos do Brasil: guia médico e biológico. Editora Roca: São Paulo; 2008.
5. Haddad V Jr, Lupi O, Lonza JP, Tyring SK. Tropical dermatology: marine and aquatic dermatology. J Am Acad Dermatol. 2009;61(5):733–50.
6. Burke WA. Coastal and marine dermatology. Presented at the Forum Meeting of the American Academy of Dermatology, San Francisco, 1997.
7. Halstead BW, Auerbach PS, Campbell DA. A colour atlas of dangerous marine animals. London: Wolfe Medical. 194 pp.; 1990.
8. Fisher AA. Atlas of aquatic dermatology. New York: Grume and Straton; 1978. 113 pp.
9. Meier J, White J. Clinical toxicology of animal venomous and poisonous. CRS: Boca Raton; 1995. 504 pp.
10. Haddad V Jr, Cardoso JLC, Neto DG. Injuries by marine and freshwater stingrays: history, clinical aspects of the envenomations and current status of a neglected problem in Brazil. J Venomous Anim Toxins Incl Trop Dis. 2014;19(1):16.
11. Haddad V Jr, Neto DG, Barbaro K, Paula Neto JB, Marques FPL. Freshwater stingrays: study of epidemiologic, clinic and therapeutic aspects based in 84 envenomings in human and some enzymatic activities of the venom. Toxicon. 2004;43:287–94.
12. Barbaro KC, Lira MS, Malta MB, Soares SL, Neto DG, Cardoso JLC, Haddad V Jr. Comparative study on extracts from the tissue covering the stingers of freshwater (Potomotrygon falkneri) and marine (Dasyatis guttata) stingrays. Toxicon. 2007;50:676–87.
13. Pedroso CM, Jared C, Charvet-Almeida P, Almeida MP, Neto DG, Lira MS, Haddad V Jr, Barbaro KC, Antoniazzi MM. Morphological characterization of the venom secretory epidermal cells in the stinger of marine and freshwater stingrays. Toxicon. 2007;50:688–97.
14. Neto DG, Haddad V Jr. Arraias em rios da região sudeste do Brasil: locais de ocorrência e impactos sobre a população. Rev Soc Bras Med Trop. 2010;43:82–8.
15. Abati PAM, Torrez PPQ, França FOS, Tozzi FL, Guerreiro FMB, Santos SAT, Oliveira SMS, Haddad V Jr. Injuries caused by freshwater stingrays in the Tapajós River Basin: a clinical and sociodemographic study. Rev Soc Bras Med Trop. 2017;50:374–8.
16. Hazin FHV, Burgess GW, Carvalho FC. Shark attack outbreak off Recife, Pernambuco, Brazil: 1992–2006. Bull Mar Sci. 2008;82(2):199–212.
17. Lentz AK, Burgess GH, Perrin K, Brown JA, Mozingo DW, Lottenberg MD. Mortality and management of 96 shark attacks and development of a shark bite severity scoring system. Am Surg. 2010;76:101–6.
18. Woolgar JD, Cliff G, Nair R, Hafez H, Robbs J. Shark attack: review of 86 consecutive cases. J Trauma. 2001;50:887–91.
19. Haddad V Jr, Gadig OBF. The spiny dogfish (cação-bagre): description of an envenoming in a fisherman, with taxonomic and toxinologic comments on the Squalus gender. Toxicon. 2005;46(1):108–10.
20. Haddad V Jr, Martins IA. Frequency and gravity of human envenomations caused by marine catfish (suborder Siluroidei): a clinical and epidemiological study. Toxicon. 2006;47(8):838–43.
21. Haddad V Jr, Souza RA, Auerbach PS. Marine catfish sting causing fatal heart perforation in a fisherman. Wilderness Environ Med. 2008;19:114–8.
22. Sazima I, Zuanon J, Haddad V Jr. Puncture wounds by driftwood catfish during bucket baths: local habits of riverside people and fish natural history in Amazon. Wilderness Environ Med. 2005;16(4):204–9.
23. Haddad V Jr, Lastoria JC. Acidentes por mandijubas (mandis-amarelos): aspectos clínicos e terapêuticos. Diagnóstico & Tratamento. 2005;10(3):132–3.
24. Aquino GNR, Souza CC, Haddad V Jr, Sabino J. Injuries caused by the venomous catfish pintado and cachara (Pseudoplatystoma sp.) in fishermen of the Pantanal region in Brazil. An Acad Bras Cienc. 2016;88(3):1531–7.
25. Negreiros MMB, Yamashita S, Sardenberg T, Fávero EL Jr, Ribeiro FAH, Haddad WT Jr, Haddad V Jr. Diagnostic imaging of injuries caused by venomous and traumatogenic catfish. Rev Soc Bras Med Trop. 2016;49(4):530–3.
26. Haddad V Jr, Stolf HO, Risk JY, França FOS, Cardoso JLC. Report of 15 injuries caused by lionfish (Pterois volitans) in aquarists in Brazil: a critical assessment of the severity of envenomations. J Venomous Anim Toxins Incl Trop Dis. 2015;21:8.
27. Vine P. Red sea safety: guide to dangerous aquatic animals. London: IMMEL; 1986. 144 pp.
28. Haddad V Jr, Martins IA, Makyama HM. Injuries caused by scorpionfishes (Scorpaena plumieri Bloch, 1789 and Scorpaena brasiliensis Cuvier, 1829) in the southwestern Atlantic Ocean (Brazilian coast): epidemiologic, clinic and therapeutic aspects of 23 stings in humans. Toxicon. 2003;42:79–83.
29. Haddad V Jr, Lastoria JC. Envenenamento causado por um peixe-escorpião (Scorpaena plumieri Bloch,

1789) em um pescador: descrição de um caso e revisão sobre o tema. Clin Ter. 2004;9(1):16–8.

30. Lehmann DF, Hardy JC. Stonefish envenomation. N Engl J Med. 1993;12;329(7):510–11.

31. Boletini-Santos D, Komegae EM, Figueiredo SG, Haddad V Jr, Lopes-Ferreira M, Lima C. Systemic response induced by *Scorpaena plumieri* venom initiates acute lung injury in mice. Toxicon. 2007;51(4):589–96.

32. Haddad V Jr, Pardal PPO, Cardoso JLC, Martins IA. The venomous toadfish *Thalassophryne nattereri* (niquim or miquim): report of 43 injuries provoked in fishermen of Salinópolis (Pará State) and Aracaju (Sergipe State). Rev Inst Med Trop São Paulo. 2003;45(4):221–3.

33. Lopes-Ferreira M, Ramos AD, Martins IA, Lima C, Conceição K, Haddad V Jr. Clinical manifestations and experimental studies on the spine extract of the toadfish *Porichthys porosissimus*. Toxicon. 2014;86:28–39.

34. Haddad V Jr, Lopes-Ferreira M, Mendes AL. Hemorrhagic blisters, necrosis, and cutaneous ulcer after envenomation by the niquim toadfish. Am J Trop Med Hyg. 2019;101:476–7.

35. Haddad V Jr, Barreiros JP. Bite by moray eel. J Venomous Anim Toxins Incl Trop Dis. 2008;14:541–5.

36. Gonçalves LF, Martins IA, Haddad V Jr. Needlefish injury in a surfer: a risk to those practicing water sports. Wilderness Environ Med. 2020;31(3):376–8.

37. Haddad V Jr, Figueiredo JL. Attack upon a bather by a swordfish: a case report. Wilderness Environ Med. 2009;20:344–6.

38. Sazima I, Guimarães SA. Scavenging on human corpses as a source for stories about man-eating piranhas. Environ Biol Fish. 1987;12:237–40.

39. Haddad V Jr, Sazima I. Piranhas attacks in southeast of Brazil: epidemiology, natural history and clinical treatment with description of a bite outbreak. Wilderness Environ Med. 2003;14(4):249–54.

40. Haddad V Jr, Sazima I. Piranhas attacks in dammed streams used for human recreation in the state of São Paulo, Brazil. Rev Soc Bras Med Trop. 2010;43:596–8.

41. Haddad V Jr, Zuanon J, Sazima I. Medical importance of candiru catfishes in Brazil: a brief essay. Rev Soc Bras Med Trop. 2021;54:e0540.

42. Jennings Simões EL. Forensic use of the piracatinga fish (*Callophysius macropterus*) to locate and identify human remains retrieved from the Amazon River. J Forensic Sci. 2018;63:1587–91.

43. Haddad V Jr, Fonseca WC. A fatal attack on a child by a black caiman (*Melanosuchus niger*). Wilderness Environ Med. 2011;22(1):62–4.

44. Campos Neto M, Haddad V Jr, Magalhães CA, Vieira IA. Attack by alligator occurred in the fisherman in the Pantanal of Mato Grosso State (Brazil): report of a case. Int J Legal Med. 2012;126: S379.

45. Reid HA. Epidemiology of sea-snake bites. J Trop Med Hyg. 1975;78(5):106–13.

46. Culotta WA, Pickwell GV. The venomous sea snakes: a comprehensive bibliography. Malabar: Krieger; 1993. 504 pp.

47. Rasmussen AR. Systematics of sea snakes: a critical review. Symp Zool Soc Lond. 1997;70:15–30.

48. Food and Agriculture Organization of the United Nations. Sea snakes. Rome: United Nations; 2007. Available at ftp://ftp.fao.org/docrep/fao/009/y0870e/y0870e65.pdf.

49. Warrell DA. Snakebites in Central and South America: epidemiology, clinical features, and clinical management. In: Campbell JA, Lamar WW (eds). The venomous reptiles of the Western hemisphere. London: Comstock, 2004. 976 pp.

50. Solórzano A. A case of human bite by the pelagic sea snake, *Pelamis platurus* (Serpentes: Hydrophiidae). Rev Biol Trop. 1995;43(1/3):321–2.

51. Sheehy CM, Solórzano A, Pfaller JB, Lillywhite HB. Preliminary insights into the phylogeography of the yellow-bellied sea snake. Pelamis platurus. Integr Comp Biol. 2012;52(2):321–30.

52. Baxter EH, Gallichio HA. Cross-neutralization by tiger snake (*Notechis scutatus*) antivenene and sea snake (*Enhydrina schistosa*) antivenene against several sea snake venoms. Toxicon. 1974;12(3):273–8.

53. Haddad V Jr, Martins IA. Kambô, an Amazonian enigma. J Venom Res. 2020;10:13–7.

54. Buriham S. Bather killed by dolphin in SP [in Portuguese]. Folha de S. Paulo Brasil. 1994 Dec 9. http://www1.folha.uol.com.br/fsp/1994/12/09/brasil/33.html.

55. Haddad V Jr. Cutaneous infections and injuries caused by traumatic and venomous animals which occurred in domestic and commercial aquariums in Brazil: a study of 18 cases and an overview of the theme. An Bras Dermatol. 2004;79(2):157–67.

56. Haddad V Jr, Miot HA, Camargo RMP, Chiaro A. Cutaneous sporotrichosis associated with a puncture in dorsal fin of a fish (*Tilapia* sp): report of a case. Med Micol. 2002;40:425–7.

57. Leme FCO, Negreiros MMB, Koga FA, Bosco SMG, Bagagli E, Haddad V Jr. Evaluation of pathogenic fungi occurrence in traumatogenic structures of freshwater fish. Rev Soc Bras Med Trop. 2011;44:182–5.

58. Neto DG, Cordeiro R, Haddad V Jr. Acidentes do trabalho em pescadores artesanais da região do Médio Rio Araguaia, Tocantins. Brasil. Cad Saúde Pública. 2005;21(3):795–803.

59. Haddad V Jr, Fávero EL Jr, Ribeiro FAH, Ancheschi BC, Castro GIP, Martins RC, Pazuelo GB, Fujii JR, Vieira RB, Neto DG. Trauma and envenoming caused by stingrays and other fish in a fishing community in Pontal do Paranapanema, state of São Paulo, Brazil: epidemiology, clinical aspects, and therapeutic

and preventive measures. Rev Soc Bras Med Trop. 2012;45(2):238–42.

60. Reckziegel GC, Dourado FS, Garrone Neto D, Haddad V Jr. Injuries caused by aquatic animals in Brazil: an analysis of the data present in the information system for notifiable diseases. Rev Soc Bras Med Trop. 2015;48(4):460–7.

61. Edilson AD, De Souza CC, Gonzales EG, Haddad V Jr, Sabino J. Avaliação do acesso a informações sobre a prevenção de acidentes por animais aquáticos cole-tados por pescadores da Bacia do Alto Paraguai, Mato Grosso do Sul. Ciênc Human Educ. 2016;16(5):460–5.

62. Macedo AKS, Silva JRP, Oliveira SP, Haddad V Jr, Vendell AL. Potentially dangerous fish of the Paraiba Estuary: identification and envenomation mechanisms. J Coast Life Med. 2017;5(11):459–62.

63. Haddad V Jr. Profile of skin diseases in a community of fishermen in the northern coast of the state of São Paulo: the expected and the unusual. An Bras Dermatol. 2019;94:24–8.

Ingestion of Venomous Aquatic Animals: Toxinology, Clinical Aspects, and Treatment

4

The earliest occurrences of human poisoning after consumption of mussels were reported in California in 1927: Sommer et al. reported the first cases of poisoning and death of mussel consumers, which are associated with the presence of seawater microalgae *Alexandrium catenella* [1, 2]. It was soon noticed that dinoflagellates that were in close associations with mussels caused poisoning in consumers. Poisonings caused by microalgae and dinoflagellates show annual outbreaks in many countries (in the USA: 30–50 cases per year). These outbreaks are concentrated in the warm months because heat favors the proliferation of dinoflagellates.

In other countries, there is little information about these poisonings: in Brazil, for example, there is little information about the outbreaks and their severity, but researchers have identified various toxins such as tetrodotoxin, microcystins, okadaic acid, palytoxin and similar, saxitoxins, and domoic acid. These toxins come from various microalgae identified as *Gambierdiscus toxicus, Microcystis aeruginosa, Dinophysis acuminata, Ostreopsis ovata, Alexandrium tamarense, Gymnodinium catenatum,* and *Pseudonitzschia* sp [3].

Teixeira et al. (1993) demonstrated the correlation between cyanobacterial blooms in Itaparica reservoir (Bahia State) and the death of 88 of the 200 people intoxicated by the reservoir water consumption between March and April 1988 [4]. In early 1996, 130 chronic renal failure patients

on hemodialysis at a clinic in Caruaru town (Pernambuco State, Brazil) started to present with serious hepatotoxicosis. Sixty of them died within 10 months after the onset of symptoms. Drinking water treatment with activated carbon used in water purification system showed the presence of cylindrospermopsins (cyanotoxins) and microcystins in the blood and liver of the affected patients [6, 7].

Envenomations by Pufferfish: Tetrodotoxin

Tetrodotoxin (Ttx) is one of the most potent toxins in the nature. It is heat stable, odorless, and colorless, and its violent neurotoxic effect is used as a defense for some animals found in different environments, including terrestrial animals. The action of the toxin is based on blocking sodium channels that are essential for the conduction of electric stimulation along the nerves [7].

Tetrodotoxin (Ttx) is probably produced by bacteria (*Vibrio* sp., *Pseudomonas* sp., *Photobacterium phosphoreum*), and it can be

V. Haddad Junior, *Medical Emergencies Caused by Aquatic Animals*,
https://doi.org/10.1007/978-3-030-72250-0_4

accumulated by animals such as pufferfish or blowfish (Diodontidae and Tetraodontidae families) (Figs. 4.1 and 4.2), some species of parrot fish and angel fish, several crustaceans (especially crabs in the Indian and Pacific Oceans), the blue-ringed octopus (genus *Hapalochlaena*) (Fig. 4.3), and some salamanders and small tropical frogs of the Dendrobatidae family (poison dart frogs) (Fig. 4.4) [7].

The fish of the Tetraodontidae family are commonly associated with poisoning by tetrodotoxin (Ttx), since they can accumulate the toxin that is produced by bacteria. Other animals may have the same defense mechanism, but the Ttx poisonings are most often caused by the consumption of pufferfish or *fugus* (Fig. 4.5). The *Takifugu, Arothron,* and *Sphoeroides* genera are those that cause accidents more often. Several pufferfish may have Ttx in their tissues, but some genera are more toxic and present greater amount of toxin.

The main toxic species of the *Takifugu* and *Arothron* genera are found in the West Indies, Japan, China, the Philippines, Taiwan, and Southern Asia,. In the New World, the main spe-

cies that cause poisoning are of the *Sphoeroides* genus: *Sphoeroides spengleri, S. testudineus, S. greeleye,* and *S. dorsalis* (spotted pufferfish) (Fig. 4.6a, b). There are toxic species of pufferfish in freshwater environments, such as the *Colomesus psittacus* and *C. asellus*: mamaiacus or freshwater pufferfish. Found in estuarine areas (*C. psittacus*) and freshwater environments (*C. asellus*), the latter is considered ornamental and is found in the Amazon basin. They are part of the family Tetrodontidae. The Diodontidae family contains the traumatic species *Chilomycterus scoepfi, C. atinga, C. antennatus,* and *C. antillarum* (the small porcupine pufferfish). The great porcupine pufferfish belongs to the *Diodon* genus: *D. histrix* and *D. holacantus.*

The pufferfish stock Ttx in the ovaries, liver, skin, and muscles. The main genera of pufferfish in the New World are *Lagocephalus* and *Sphoeroides,* but the capacity of intoxication is different between them: Oliveira, in his master's thesis, showed that tetrodotoxin levels in the muscles, skin, and guts of the *Sphoeroides* genus are high and represent risk to consumers, while

Fig. 4.1 *Chylomycterus* sp, a porcupinefish of the Diodontidae family. (Photo: Vidal Haddad Junior)

Fig. 4.2 The most responsible for human envenomations in the New World is the genus *Sphoeroides*, a pufferfish common in shallow waters. (Photo: Vidal Haddad Junior)

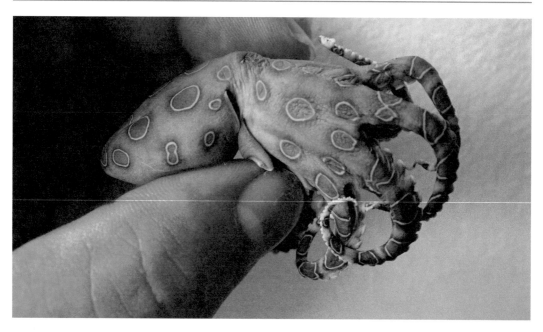

Fig. 4.3 Maculotoxin, a neurotoxin present in the blue-ringed octopus (*Hapalochlaena* sp.) is identical to tetrodotoxin. (Photo: Vidal Haddad Junior)

Fig. 4.4 Presence of poisons in tropical frogs is frequent. Various neurotoxins can be found in these animals, including tetrodotoxin and analoges. In the image: *Dendrobates azureus*, the blue dart frog. (Photo: Vidal Haddad Junior)

Fig. 4.5 Fugu is the Japanese name to pufferfish (*Takifugu* genus). There are various poisonous species in the genus

Fig. 4.6 (**a**) *Sphoeroides* sp., the spotted pufferfish. (Photo: Vidal Haddad Junior). (**b**) The sale of spotted pufferfish (*baiacu-pinima*) occurs in several fish markets in Brazil in Northeast Coast of Brazil, even with the risks of ingestion. According to the fishermen, it is enough to remove the viscera so that there is no more danger of poisoning, which does not correspond to reality. (Photos: Vidal Haddad Junior)

the levels of Ttx in the *Lagocephalus* genus (Fig. 4.7) are lower and rarely present risks to humans in the area of the study (Brazil). Pufferfish can also provoke severe traumatic lesions in the hands of fishermen, caused by the oral plaques, which are not really tooth (Fig. 4.8).

Clinical Aspects

In the first 5–45 minutes after consuming the fish, there appears a sensation of "numbness," perioral paresthesias (an initial symptom, typical of poisoning by neurotoxins), nausea, and vomiting. In 10–60 minutes, speech becomes impaired due to anesthesia and motor blockade of the tongue and face muscles.

In a period of 1–6 hours, progressive muscular paralysis with arterial hypotension and respiratory failure is noted. The pupils can be dilated and fixed and the patient presents the "prison" syndrome, which is a progressive and complete muscle paralysis with awareness maintained. Within 24 hours, total respiratory paralysis and severe cardiac arrhythmias are possible. Death can occur in this phase. If the patient exceeds this critical period, there is a tendency to recover without sequelae in about 48 hours [8–10].

In a series of 27 envenomations caused by pufferfish of the *Sphoeroides* genus in Bahia and Santa Catarina States (Brazil), the clinical manifestations were vomiting in 13 patients (48%), paresthesias in 12 (44%), nausea in 9 (41%), dizziness, abdominal pain, dyspnea, respiratory failure and muscular paralysis in 6 (22%), muscle weakness, fasciculation, and ataxia in 4 (15%), malaise, cardiopulmonary arrest, and coma in 3 (11%), arterial hypotension in 2 (7%), and aphonia in 1 patient (4%). Two patients died (7%) [8].

Treatment

- Ventilatory support: fundamental, especially for high mortality when installing respiratory muscle paralysis.
- Gastric lavage: up to 3 hours after ingestion.
- Bicarbonate of soda: appears to reduce the toxicity of the venom.
- Some clinical studies point that fampridine (4-aminopyridine) reverses the effects of tetrodotoxin and saxitoxin poisoning in experimental studies, but its effectiveness in humans has not yet been determined [11].

Fig. 4.7 *Lagocephalus lavegatus*, the macaw pufferfish. (Photo: Vidal Haddad Junior)

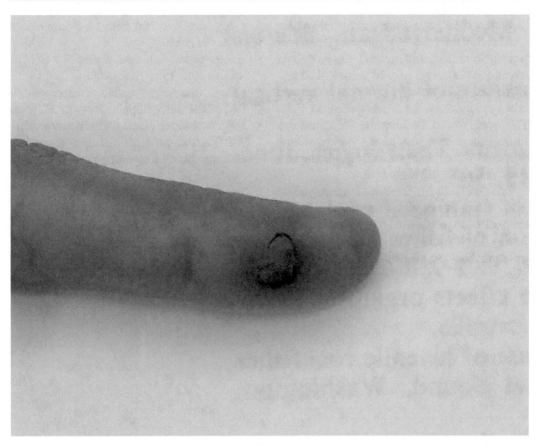

Fig. 4.8 This patient was bitten by a pufferfish when collecting fish for a research. Note the marks of the oral plaques and the format of the wound. (Photo: Vidal Haddad Junior)

Current data: Tokyo Bureau of Social Welfare and Public Health: 20–44 poisonings per year from 1996 to 2006 (6% deaths). Only one was reported in restaurant (the fish "unload" their toxicity when taken out of their environment and raised in aquariums for consumption). In 1958, 176 people died. Data from the Fugu Research Institute show that 50% of the victims ate the liver and 7% the skin of the fish. Japan's emperor cannot eat puffer fish meat, by law.

In 2007 (Thailand), there was a warning about puffer fish meat sale, like salmon, which caused the death of 15 people and 115 hospitalizations in 3 years. In November 2011, a two-star-Michelin restaurant boss lost his job for serving fugu liver to a customer who had been later hospitalized with risks of death.

- Foxnews.com: Poisonous Puffer Fish Sold as Salmon Kills 15 in Thailand
- Father dies after eating puffer fish – INQUIRER.net, Philippine News for Filipinos
- http://www.dailymail.co.uk/news/article-2070361/Two-Michelin-star-chef-suspended-customer-nearly-dies-eating-puffer-fish.html

The serpent and the rainbow – Wade Davis: Research involving the link among pufferfish venom of the *Sphoeroides* genus and voodoo cults and zombies legends in Haiti.

Box 4.1 Pufferfish

A male child, 1 year and 11 months old, was admitted to the Goiana Town Hospital, Pernambuco State (Brazil), with a history of ingestion of pufferfish viscera. About 1 hour after the meal and 2 hours before being admitted to the hospital, there was an onset of symptoms of poisoning in the child.

The family members said that they caught some specimens of pufferfish the previous night of the accident on an estuary located in a district in Goiana (Figs. 4.9 and 4.10). The boy's grandmother cleaned the fish, removing the skin and the viscera, which were washed in running water and soaked with lemon juice. Only the viscera were prepared, and they were fried in hot oil and then mixed with sweet potatoes. The child ingested the preparation in the morning, and according to the family, approximately 1 hour after the meal, the child had cold sweating, malaise, and softening of the body. Two hours after eating the viscera, the child developed intense salivation, cyanosis, and cardiopulmonary arrest (CPA), and he was taken to the nearest health facility by the family members. Resuscitation maneuvers were instituted without any success. The fish was identified as the spotted pufferfish of the *Sphoeroides* genus.

Commentaries: the spotted pufferfish (*Sphoeroides* genus) is the main agent of severe envenomations in the Atlantic Ocean. The consumption of viscera (including the liver) aggravates the poisoning.

Fig. 4.9 The spotted pufferfish is responsible for tetrodotoxin envenomation in America. (Photo: Vidal Haddad Junior)

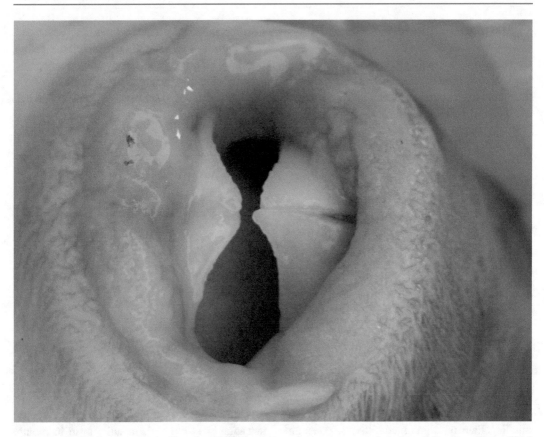

Fig. 4.10 The oral plaques of the Tetrodontidae fish can provoke severe traumatic lesions in humans. (Photo: Vidal Haddad Junior)

Ciguatoxins: Ciguatera

Ciguatoxin is a toxin produced by dinoflagellates, and the species *Gambierdiscus toxicus* is the main incriminated species. The name "cigua" comes from a gastropod mollusk founded in the Caribbean, where the disease has been described. It is harmless to fish. The toxin is heat stable, odorless, colorless, and acts on sodium channels, thereby blocking muscular activities. Maitotoxin can cause similar manifestations but acts on calcium channels.

Ciguatoxin poisoning occurs by ingestion of large reef fish, which accumulate larger amounts of ciguatoxin. The chain begins with the production of ciguatoxin by the dinoflagellate *G. toxicus*, followed by the establishment of these dinoflagellate on algae macrophytes, consumption of algae by herbivorous fish, and predation of the herbivorous by carnivorous fish (highly toxic concentrations). The main fish associated with ciguatera are groupers (*Mycteroperca* and *Epinelephus* genera) and barracudas (*Sphyraena* genus), and more than about 50,000 cases occur per year mainly in the Caribbean. But the disease is present in all tropical regions [12–17].

Clinical Aspects

After the consumption of contaminated meat, nausea, vomiting, diarrhea, muscle pain, and weakness are observed in 10 minutes to 36 hours. Perioral paresthesia is again present and also paresthesias in extremities arises (IMPORTANT: suspect of neurotoxins!).

Other signs and symptoms are chills, intense sweating (wet clothes), metallic taste in mouth, hypotension, bradycardia, muscular pain/paralysis, difficulty in breathing, cyanosis, cardiac arrhythmias, and respiratory failure but with low risk of death. Ciguatera poisoning presents special characteristics compared with other poisonings by neurotoxins. Paradoxical thermal sense is very characteristic: the patient reverses the hot/cold senses of water temperature (occurs about 2 days after consumption). A severe itch starts in the palms and then becomes widespread. It may be recurrent after consumption of fish, seafood, nuts, or alcohol for long periods. Neurological symptoms may persist for years: mainly involuntary movements and muscle weakness occur [12–17].

Box 4.2 Ciguatera

A 30-year-old man who was on his vacation in Cancun traveled to Isla de las Mujeres, Mexico, where he ate barracuda fish for lunch (Fig. 4.11). Two and a half hours after having the meal, he noticed oral paresthesia that progressed to paresthesia and weakness on the legs and finally on the whole body. Liquid diarrhea followed and, fearing having eaten spoiled food, he forced vomiting (a dark liquid came out). The diarrhea lasted for 3 days.

On the second day, the patient had hiccups that were only controlled by drugs. On the first day, he complained of metallic taste and a strong odor in his body. During the 2 days of hospital stay, he received intravenous mannitol. He also had a generalized pruritus (which began on the fourth day after exposure) and dysesthesia, having a ice-burning feeling when touching the water at room temperature. Twenty days after exposure, he was using hydroxyzine 25 mg 6 times a day, gabapentin 300 mg 6 times a day, and dexamethasone 4 mg 2 times a day, but he still had pruritus (especially when exposed to cold) and could not sleep at night. Only 23 days after exposure, he got to sleep normally and 28 days after exposure the pruritus remitted, but he still had burning sensations on the legs and exacerbated sensations to cold. Bradycardia lasted for a couple of days. Metallic taste and a strong odor were also his complaints during the initial days of exposure. Laboratory exams were normal.

Commentaries: this case reports a classic ciguatera poisoning with problems and typical clinical manifestations associated with the ingestion of barracuda fish.

Fig. 4.11 The consumption of the meat of barracudas (*Sphyraena* genus) are frequently associated with ciguatera envenomation, especially in the Caribbean. (Photo: Vidal Haddad Junior)

Treatment

- THERE IS NO ANTIDOTE!!
- Gastric lavage (up to 3 hours; the toxin is absorbed after this period).
- Mannitol 20% can be useful: the dosage is 1 g/kg EV in 30 minutes.
- Prolonged neurological symptoms need symptomatic medication.

Seafood Paralysis (Paralytic Shellfish Poisoning) and Red Tide: Saxitoxin and Gonyautoxins

Both entities are caused by saxitoxin and gonyautoxins and derivatives. These neurotoxins are absorbed by filter feeder shellfish, especially mussels and oysters, being produced by dinoflagellates of the genera *Alexandrium*, *Gonyaulax*, *Gimnodinium*, and *Pyrodinium*.

Saxitoxin and gonyautoxin act by blocking sodium channels, thereby inhibiting muscle and causing gastrointestinal and neurological manifestations. Paralysis by seafood occurs when the victim consumes contaminated shellfish. Overpopulation of dinoflagellates causes red tides, which results in massive death of marine fauna. This intoxicates humans and causes breathing problems with nasal discharge that affects the victims near the water. In the Bible, there is the sentence: "... And the waters became blood. The river became smelly and the fish died. The Egyptians could not drink the river water (Exodus, Chapter 7, Verses 20-21)" (Fig. 4.12) [12, 19–20].

Fig. 4.12 A *Noctiluca* bloom in Union Bay, British Columbia. *Noctiluca* sp. is a non-toxic species, but the bloom shows the impressive aspect of a red tide. (Photo: Lisa M. Holm, BC)

Clinical Aspects

In 5–30 minutes, perioral paresthesia occurs (IMPORTANT SYMPTOM!), extending to the face and neck. Within hours, it is possible to observe nausea, vomiting, abdominal pain, diarrhea, dysarthria, muscle weakness, malaise, tingling and numbness of extremities, difficulty in breathing, and severe muscle paralysis. There may be temporary blindness, a feeling of floating in space, and intense thirst. Muscle paralysis can lead to death (rare).

Treatment

- *There Is No Antidote!*
- Respiratory support treatment and gastrointestinal treatment are important, as gastric lavage occurs in the first 3 hours.

Daniel Kohane, Children's Hospital Harvard Medical School in Boston, Massachusetts, tested in animals a slow-release system via liposomes (miniblisters or minicells) successfully. The study was published in April 2009 in the Proceedings of the National Academy of Sciences.

Brevetoxins: Neurotoxicity by Seafood

Brevetoxins are neurotoxins that bind to sodium channels in cell membranes, causing persistent activation of nerve cells, skeletal muscle, and cardiac cells. Several dinoflagellates produce brevetoxins, the most important genus being *Gymnodium* (currently *Karenia brevis*) [21–23].

Neurotoxicity by seafood (neurotoxic shellfish poisoning or NSP) is caused by the consump-

tion of filter feeder shellfish (clams, mussels, oysters) contaminated with brevetoxins.

Clinical Aspects of NSP

Clinical aspects of NSP are paresthesias (initially in perioral region but also in extremities), reverse temperature, myalgia, dizziness, ataxia, abdominal pain, nausea, vomiting, headache, bradycardia, and dilated pupils [21–23].

Treatment

- There Is No Antidote!
- Respiratory support and gastrointestinal treatment are important.
- Gastric lavage can be useful in the initial hours.

Amnesia by Seafood Toxins: Domoic Acid

Amnesia by seafood toxins is caused by diatoms of the genus *Pseudo-nitzschia,* and its transmission is by eating sardines, anchovies, and seafood and can victimize seabirds and mammals. Domoic acid is a toxin, which acts as an antagonist of glutamate, a neurotransmitter in the central nervous system. This acid is a neuroexcitatory amino acid, which enhances the action of natural excitatory amino acids such as glutamate. It acts on glutamate receptors in the central nervous system, inducing depolarization of the postsynaptic membrane [24–25].

Clinical Aspects

Within 24 hours, nausea, vomiting, and diarrhea occur. Within 48 hours, the victim presents with neurological symptoms: headache, mental confusion, behavioral changes, and memory loss, which can be severe. In elderly patients, brain injury, coma, and death can occur. The memory loss resembles the manifestations of Alzheimer's disease, and there was an association between memory loss and age: patients under 40 years predominantly had diarrhea and those with more than 50 years had memory loss [24–25].

In the 1960s, thousands of sea birds (especially seagulls) began to behave in an anarchic way in the Santa Cruz and Capitolia regions, California, bumping on windows, homes, and other places, and a large number of birds were found dead and some regurgitating anchovies. The episode was reported by local newspapers and had great repercussion, leading the filmmaker Alfred Hitchcock to create the plot of the movie "The Birds," a classic thriller of suspense. In 2011, scientists at the University of Louisiana studied plankton collected in the region at that time and found large quantities of *Pseudo-nitzschia* algae, which produces domoic acid and was likely responsible for the change in the behavior of birds, a fact previously described [26].

Treatment

- There Is No Antidote!
- Neurological support treatment is important, as so the gastric lavage.

Diarrhea by Seafood: Okadaic Acid

Several dinoflagellates, particularly of the *Dinophysis* genus, produce okadaic acid and contaminate seafood, especially mussels [27–29].

Bivalve molluscs: Santa Catarina State, Brazil, January 2007: more than 150 cases of "Diarrhoeic Shellfish Poisoning" (DSP) associated with the alga *Dinophysis acuminata* and okadaic acid were reported, which were confirmed by specific bioassays ("mouse bioassay") (Figs. 4.13 and 4.14).

The mollusk incriminated was the mussel (*Perna perna*) grown in that state. Subsequently, (Fig. 4.15) [30].

Clinical Aspects

The signs and symptoms start to occur from 30 minutes to 12 hours (about 4 hours), with gastrointestinal symptoms such as vomiting, nausea, and importantly, diarrhea. The manifestations persist for around 4 days and can be confused with bacterial infections, but no fever occurs [27–30].

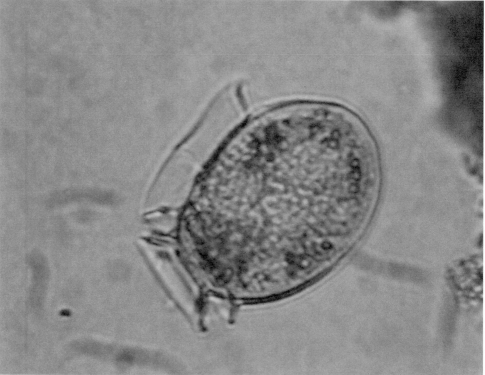

Figs. 4.13 and 4.14 *Dinophysis* cf *acuminata* bloom in the Canto Grande beach, Bombinhas town, Santa Catarina, Brazil, in August 2008. The event interrupted the harvest and consumption of shellfish in the region due to the presence of okadaic acid above the normal recommended levels for human consumption. (Courtesy of Luis Antonio Proença, Laboratory of Research on Nocive Algae and Ficotoxins – IFSC/Itajaí, Santa Catarina State, Brazil)

Fig. 4.15 A great offer of mollusks in fish and seafood markets can improve the risks of envenomations. (Photo: Vidal Haddad Junior)

Treatment

- Symptomatic (may require hospitalization).

Poisoning After Consumption of Octopus (*Octopus* sp.)

Haddad Jr. & Moura (2007) reported an envenomation manifested by neurological and muscular symptoms in a woman of 45 years that occurred after the consumption of raw meat of common octopus (*Octopus* sp.) (Fig. 4.16) [31].

Clinical Aspects

The presence of neuromuscular symptoms suggests the action of neurotoxin, a fact that proves that many kinds of common octopus contain cephalotoxins. Very little is known about the toxins of the genus *Octopus* [32].

Scombroidism

The meat of the fish of the Scombroidea family deteriorates rapidly. The tuna, for example, are caught and immediately thrown into ice chambers (Fig. 4.17). In such fish, there are high levels of histidine in the tissues and rapid deterioration gives rise to saurine and histamine. Saurine, the main toxin involved in this type of poisoning, is formed by the action of bacteria (especially *Proteus morganii*) on the poorly maintained meat of fish with high blood supply, such as tuna, bonito, and mackerel (*Scombroides* gender – Fig. 4.17). The disease presents with the signs and symptoms of histamine poisoning [33–35].

Clinical Aspects

Within 30–60 minutes, pulsatile headache, vomiting, diarrhea, tachycardia, dry mouth, edema and erythema of the face, conjunctivitis, blisters

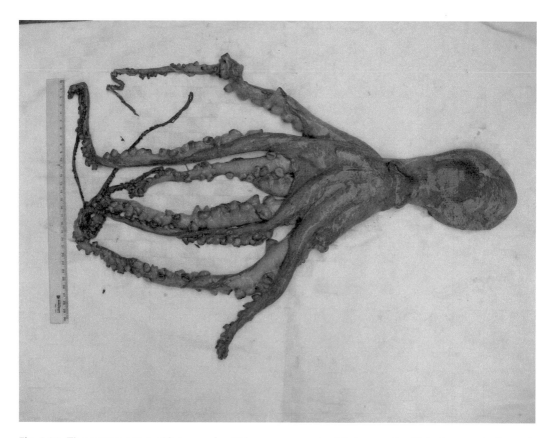

Fig. 4.16 The common octopus (*Octopus vulgaris*) is a venomous animal, being capable of injecting neurotoxins, such as cephalotoxin. (Photo: Vidal Haddad Junior)

Fig. 4.17 The fish of the Clupeoformes order include sardines (one of the most consumed fish in the world), anchovies, herring, and tarpons. The toxin in these fish has not been identified, but it is suspected to be accumulation of any substance present in dinoflagellates. (Photo: Vidal Haddad Junior)

on the trunk and face, difficulty in breathing, and death (rare) occur. Recovery begins within 24 hours [33–35].

In April 2016, an outbreak of scombroidism was recorded in the Archipelago of Fernando de Noronha, in the Northeast of Brazil. Thirty-one people presented with cutaneous erythema, perioral paresthesias, hives, and vertigo, about half an hour after consuming yellowfin meat or albacora (*Thunnus alalunga*), which belongs to the family Scombridae, due to the deterioration of histidine. The fish showed high levels of histamine, proving the initial hypothesis. The mandatory conservation of fish immediately on ice solved the problem, and the problem has not been reported again. (http://g1.globo.com/pernambuco/blog/viver-noronha/post/proibida-pesca-sem-uso-de-gelo-em-fernando-de-noronha.html).

Treatment

• Antihistamines show good results.

Clupeotoxicity

The fish of the order Clupeoformes include sardines (one of the most consumed fish in the world), anchovies, herring, and tarpons (Fig. 4.18). The toxin in these fish has not been identified, but it is suspected to be accumulation of any substance present in dinoflagellates but different from ciguatoxin, because the fish involved feed on plankton, unlike the carnivorous fish that are associated with ciguatera [12].

Fig. 4.18 Scombroid poisoning occurs due to the rapid degradation of histidine into histamine in Scombroidae family of fish such as tuna (*Thunnus* genus)

Clinical Aspects

Envenomations are rare and were described in the Indian and Pacific Oceans and the Caribbean. The symptoms begin in 15–90 minutes, with dryness and metallic taste in the mouth (characteristic of the disease), nausea, vomiting, and diarrhea. Severe cases show severe compromise of the central nervous system, manifested through dilated pupils, headache, numbness, tingling, intense salivation, cramps, difficulty in breathing, muscle paralysis, convulsions, coma, and death. The mortality rate has reached 45%, and some deaths happened so quickly that the fish was still found in the victim's mouth at the time of the death [12].

Poisonous Sharks

Under certain conditions, the meat and the liver of some sharks, such as the bull shark (*Carcharinus leucas*) and the tiger shark (*Galeocerdo cuvieri*), can become toxic (Fig. 4.19).

Clinical Aspects

Intoxication is manifested through neurological symptoms such as seizures, breathing difficulties, coma, and death. Ataxia was the main aspect of the intoxication. In an outbreak observed in southeast coast of Madagascar, 68 victims died after feeding on the flesh of a single bull shark. The mortality rate reached 30%. The toxins involved were called carchatoxins A and B, but there is no further information on its composition or origin [36].

Poisoning by Other Fish

Crinotoxic fish are those with toxicity in secretions present on their skin. Envenomations by these fish in humans are rare and fish use this defense against predators. The soapfish of family Ostraciontidae produce pahutoxina, and some flounders have in their skin pardaxin and pavonins, powerful repellents of predators.

Fig. 4.19 The tiger shark (*Galeocerdo cuvier*) is one of the species associated with envenomations by consumption of shark's meat. (Photo: Vidal Haddad Junior)

Palytoxin is one of the most potent toxins present in the nature. Intoxication occurs through the consumption of meat of fish of the Balistidae family (triggerfish) (Fig. 4.20). These feed on coral zoanthids (*Palythoa* genus). Palytoxin was originally isolated in these corals, but it appears to be produced by bacteria and accumulated in some animals, with a supposed mechanism similar to the accumulation of tetrodotoxin in some animals [37, 38]. Some authors associate palytoxin with ciguatera, clupeotoxism, and Haff disease.

Clinical Aspects

Envenomation causes rhabdomyolysis, metallic taste in the mouth, abdominal cramps, nausea, vomiting, diarrhea, lethargy, paresthesia, bradycardia, renal failure, impairment of sensation, muscle spasms, tremor myalgia, cyanosis, and respiratory distress. The inhalation of aerosols can cause respiratory alterations, such as wheezing, bronchoconstriction, dyspnea, and conjunctivitis.

Fig. 4.20 Fish of the Balistidae family (triggerfish) can accumulate palyotoxin after feeding on zoanthid corals. (Photo: Vidal Haddad Junior)

Minamata Disease: Mercury Poisoning

Minamata Bay (Japan): methylmercury release by Chisso industry for years provoked teratogenicity and neurological poisoning in humans and other animals, initially perceived in the middle of the 1950s.

Clinical Aspects

The disease causes ataxia, extremity numbness, muscle weakness, and loss of visual field, audition, and speech. In extreme cases, dementia, paralysis, coma, and death weeks after the onset of the disease are possible. There is a congenital form that affects fetuses in the womb. In several places in the world, mercury is used in gold mining in rivers and, thus, there is a possibility of poisoning by this heavy metal.

The environmental impact of mercury accumulation in other countries is less, but real, compared to that of Minamata Bay. Recent checks in gold mining areas in rivers in the Amazon (many of them illegal and therefore uncontrolled) have shown high levels of methylmercury in carnivorous fish. Among these, there are siluriformes (catfish) such as dorado (*Brachyplatystoma rousseauxii*), piraiba (*Brachyplatystoma filamentosum*), surubim (*Pseudoplatystoma* sp.), and mandubés (*Ageneiosus brevifilis*) and scales fish, such as the piranhas (Serrasalmidae), the peacock bass (*Cichla* sp.), and the trairão (*Hoplias lacerdae*). In some specimens, mercury levels were four times higher than that accepted by WHO [39].

Methylmercury used to purify gold accumulates in an increasing way in the food chain, reaching maximum levels in larger carnivorous fish, similar to what occurs with dinoflagellate toxins, as the ciguatoxin. Previous studies have shown levels above the normal in the hair of riverside populations, especially those who live near the rivers where mining is done, highlighting the need for discussions about the problem.

Accumulation of mercury can cause fetal malformations and delay child development. As mercury accumulates and is not eliminated, there are possibilities of serious outbreaks of poisoning in the future in various countries of South America. The Hg values tolerated by the Brazilian Health Agencies are 1 mg/kg in carnivorous fish and 0.5 mg/kg in herbivorous fish [39].

Poisoning by Marine Turtles

One of the lesser known poisoning caused by marine animals occurs through ingestion of turtle meat. There are several outbreaks reported in the waters of the Indian and Pacific Oceans regions, especially in the Malay Archipelago, Sri Lanka, Cambodia, and southern India. Since the first proven communication of poisoning by turtles, the suspected mechanism is the accumulation of toxic substances in algae or dinoflagellates, more likely the former. This hypothesis is strengthened by the poisonings that occur in restricted areas, which could be linked to certain species of algae in that specific areas on which the turtles feed.

The main species involved in outbreaks is *Eretmochelys imbricata*, the hawksbill turtle (Fig. 4.21). However, there are reports of poisonings by *Chelonia mydas* (the green turtle) and *Dermochelys coriacea* (the leatherback turtle). Since 1987, 152 deaths were recorded in the Indo-Pacific region (155 to an outbreak reported in Cambodia in December 2002). A total of 99 deaths were caused by eating hawksbill turtle, 46 by green turtles, and 15 by leatherback turtles. Approximately 70% of people who eat contaminated meat survive and the high mortality rate is nearly 30%. There are descriptions of five deaths in children who had been fed breast milk of mothers intoxicated, which shows the power of the poison [12]. The toxin involved is called chelonitoxin and has not yet been identified. The poisoning affects the nervous system and causes massive liver necrosis.

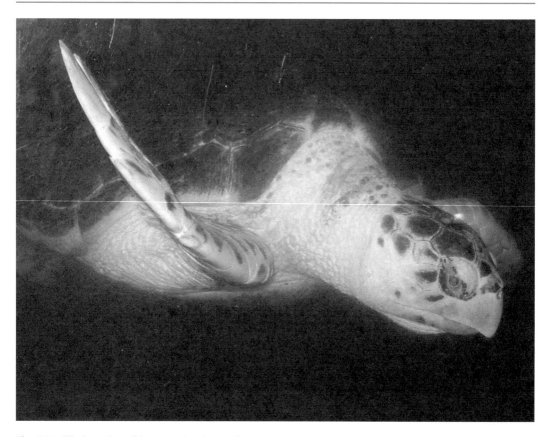

Fig. 4.21 The ingestion of the meat of marine turtles can provoke serious envenomations, possibly by toxic substances in algae or dinoflagellates devoured by the reptiles. (Photo: Vidal Haddad Junior)

Clinical Aspects

The patients presents with nausea, vomiting, tachycardia, paleness, stomach pain, cold sweats, dizziness, cold extremities, tingling around the mouth, and sores on the tongue, mouth, and palate, which are the most important signs of this type of poisoning . In more severe cases, convulsions and death occur. There is no antivenom and the treatment is symptomatic. In the region of envenomations, the fishermen scrub blood of the turtle on the skin before consuming the meat. If irritation occurs, they avoid consumption. The only possible prevention is to avoid eating the meat of any marine turtle [12].

Haff Disease

Haff disease was first described in 1924 in the city of Königsberg, in the Baltic, among people living near a pond (German: haff). The disease presented an unexplained rhabdomyolysis in individuals who fed on freshwater fish about 24 hours before the clinical manifestations.

Over the next 15 years, about 1000 cases have been reported in humans, birds, and cats during the summer and fall. A connection was made with the consumption of freshwater fish (eel, pike). Most of the cases were observed in Germany and Russia. In the USA, six cases were reported after consumption of buffalo fish (*Ictiobus cyprinellus*) in 1997.

The toxin is not known. Studies of the CDC (USA) showed a hexane-soluble substance (nonpolar lipid) which induced similar symptoms in rats. The substance is not inactivated by cooking (all the victims had eaten fried or boiled fish) – Buchholz et al. [42].

Twenty-five cases of rhabdomyolysis were identified in Manaus city, Amazon State, Brazil, between June and September 2008. All patients had consumed fried or baked pacu 24 hours before the onset of symptoms: silver pacu (*Mylossoma* spp), black-finned tambaqui (*Colossoma macropomum*), and pirapitinga or freshwater pompano (*Piaractus brachypomus*) – three omnivorous fish of the same family (Fig. 4.22) [40].

In 2016, 64 suspected cases of Haff disease were recorded in the state of Bahia in northeastern Brazil. Patients had severe myalgia, darkened urine, elevated creatine phosphokinase (CPK) and aspartate aminotransferase (AST), and malaise. All had eaten meat of the fish Olho de Boi (*Seriola lalandi*) and grouper (*Acanthistius brasilianus*). The cases were not confirmed, as it is not known which toxin causes the Haff syndrome, but the clinical and epidemiological evidences strongly pointed to the disease.

(https://www.metro1.com.br/noticias/bahia/31889,intoxicacao-apos-ingestao-de-peixe-e-causa-de-doenca-misteriosa-indica-estudo).

Clinical Aspects

Intense muscle pain, chest pain, striking rhabdomyolysis, high serum CPK.

Treatment

- Symptomatic.
- Kidney failure?

Fig. 4.22 The tambaqui or colossoma black-finned (*Colossoma macropomum*) and pirapitinga or freshwater pompano (*Piaractus brachypomus*) are the species associ- ated with cases of Haff disease in the Brazilian Amazon. (Photo: Vidal Haddad Junior)

Box 4.3 Haff Disease

A 24-year-old female patient presented to an emergency room with severe generalized muscle pain that started as low back pain, which had appeared three days earlier, when she was still on vacation in the city of Salvador, state of Bahia. She denied use of medications, animal contacts, gastrointestinal symptoms, or other problems. Approximately 1 day before the onset of the myalgias, she had consumed greater amberjack or "olho-de-boi" (*Seriola dumerilii*), a marine fish widely consumed in the region. When she returned to her home state (Rio de Janeiro), she still had intense pain and her urine progressively darkened.

Haff disease was suspected, and her creatinine phosphokinase (CPK) level was above 10,000 U/L (normal values up to 170), and aspartate aminotransferase (AST) and alanine aminotransferase (ALT) levels were found to be elevated, with both the values above 100 U/L, confirming the hypothesis of rhabdomyolysis and the presumptive diagnosis of Haff syndrome.

She was symptomatically medicated for the prevention of renal failure due to rhabdomyolysis and, therefore, she progressively improved and was discharged in approximately 10 days.

Comments: the presence of rhabdomyolysis and consumption of one of the fish associated with the appearance of rhabdomyolysis reinforced the diagnosis of Haff syndrome. The treatment was symptomatic, as the toxins involved in the disease are not known. The patient evolved well, but there are reports of multiple organ failure caused by the disease (Fig. 4.23).

Fig. 4.23 The main fish associated with Haff syndrome in freshwater environments are pacus and tambaquis (genera *Mylossoma*, *Colossoma* and *Priaractus*). In the details: the cases in marine environments are linked to the consumption of groupers (above) and, mainly, great amberjacks (below)

References

1. Sommer H, Meyer KF. Paralytic shellfish poisoning. Arch Pathol. 1937;24:560–98.

2. Schantz EJ. Historical perspective on paralytic shellfish poison. In: Seafood toxins. Washington, DC: American Chemical Society; 1984. p. 99–111.

3. Proença LAO, Mafra LL. Ocorrência de ficotoxinas na costa brasileira. In: SBFIC. (Org.), editor. Formação de Ficólogos: um compromisso com a sutentabilidade dos recursos aquáticos. Rio de Janeiro; 2005. p. 57–77.

4. Teixeira MGLC, Costa MCN, de Carvalho CLP. Gastroenteritis epidemic in the area of Itaparica dam, Bahia, Brazil. Bull Pan Am Health Organ. 1993;27:244–53.

5. Jochimsen EM, Carmichael WW, An JS, Cardo DM, Cookson ST, Holmes CE, Antunes MB, de Melo Filho DA, Lyra TM, Barreto VS, et al. Liver failure and death after exposure to microcystins at a hemodialysis center in Brazil. N Engl J Med. 1998;338(13):873–8.

6. Azevedo SM, Carmichael WW, Jochimsen EM, Rinehart KL, Lau S, Shaw GR, Eaglesham GK. Human intoxication by microcystins during renal dialysis treatment in Caruaru-Brazil. Toxicology. 2002;181/182:441–6.

7. Haddad V Jr, Takehara ET, Rodrigues DS, Lastória JC. Poisonings by pufferfish: a review. Diagnostico e Tratamento. 2004;9:183–5.

8. Silva CCP, Zannin M, Rodrigues DS, Santos CR, Correa IA, Haddad V Jr. Clinical and epidemiological study of 27 poisonings caused by ingesting puffer fish (Tetradontidae) in the states of Santa Catarina and Bahia, Brazil. Rev Inst Med Trop Sao Paulo. 2010;52(1):47–50.

9. Santana Neto PL, de Aquino ECM, Silva JA, Porto Amorim ML, Oliveira Junior AE, Haddad V Jr. Fatal poisoning by pufferfish (Tetradontidae) - a case report in a child. Rev Soc Bras Med Trop. 2010;43(1):92–4.

10. Simões EMS, Mendes TMA, Adão A, Haddad V Jr. Poisoning after ingestion of pufferfish in Brazil: report of 11 cases. J Venom Anim Toxins Incl Trop Dis. 2014;20:54.

11. Chang FC, Spriggs DL, Benton BJ, Keller SA, Capacio BR. 4-Aminopyridine reverses saxitoxin (STX) and tetrodotoxin (TTX) induced cardiorespiratory depression in chronically instrumented guinea pigs. Fundam Appl Toxicol. 1997;38(1):75–88.

12. Scott S, Thomas C. All stings considered: first aid and medical treatment of Hawaii's marine injuries. Honolulu: University of Hawaii; 1997. 248 pp.

13. Cameron J, Capra MF. The basis of the paradoxical disturbance of temperature perception in ciguatera poisoning. J Toxicol Clin Toxicol. 1993;31(4):571–9.

14. Friedman MA, Arena P, Levin B, Fleming L, Fernandez M, Weisman R, et al. Neuropsychological study of ciguatera fish poisoning: a longitudinal case-control study. Arch Clin Neuropsychol. 2007;22(4):545.

15. Friedman MA, Fleming LE, Fernandez M, Bienfang P, Schrank K, Dickey R, et al. Ciguatera fish poisoning: treatment, prevention and management. Mar Drugs. 2008;6(3):456–79.

16. Schnorf H, Taurarii M, Cundy T. Ciguatera fish poisoning: a double-blind randomized trial of mannitol therapy. Neurology. 2002;58(6):873–80.

17. Dickey RW, Plakas SM. Ciguatera: a public health perspective. Toxicon. 2010;56(2):123–36.

18. Alonso JF. Mareas rojas y biotoxinas: química y epidemiología. Santiago: Conselleria de Sanidad de la Xunta de Galicia; 1989, 1989.

19. CDC. Epidemiologic notes and reports paralytic shellfish poisoning - Massachusetts and Alask. MMWR. 1991;40(10):157–61.

20. Gessner BD, Bell P, Doucette GJ, Moczydlowski E, Poli MA, Van Dolah F, Hall S. Hypertension and identification of toxin in human urine and serum following a cluster of mussel-associated paralytic shellfish poisoning outbreaks. Toxicon. 1997;35(5):711–22.

21. Lee MS, Qin G, Nakanishi K, Zagorski MG. Biosynthetic studies of brevetoxins, potent neurotoxins produced by the dinoflagellate *Gymnodinium breve*. J Am Chem Soc. 1989;111(16):6234–41.

22. Watkins SM, Reich A, Fleming LE, Hammond R. Neurotoxic shellfish poisoning. Mar Drugs. 2008;6(3):431–55.

23. Flewelling LJ, Landsberg JH, Naar JP. Karenia brevis red tides and brevetoxin-contaminated fish: a high risk factor for Florida's scavenging shorebirds? J Bot Mar. 2012;55(1):31–7.

24. Bates SS, Bird CJ, de Freitas ASW, Foxall R, Gilgan M, Todd ECD, et al. Pennate diatom *Nitzschia pungens* as the primary source of domoic acid, a toxin in shellfish from eastern Prince Edward Island, Canada. Can J Fish Aquat Sci. 1989;46:1203–15.

25. Todd ECD. Domoic acid and amnesic shellfish poisoning - a review. J Food Prot. 1993;56(1):69–83.

26. Bargus S, Silver MW, Ohman MD, Benitez-Nelson CR, Garrison DL. Mystery behind Hitchcock's birds. Nat Geosci. 2013;5:2–3.

27. Yasumoto T, Oshima Y, Sugawara W, Fukuyo Y, Oguri H, Igarashi T, Fujita N. Identification of *Dinophysis fortii* as the causative organism of diarrhetic shellfish poisoning. Bull Jpn Soc Sci Fish. 1980;46(11):1405–11.

28. Murata M, Shimatani M, Sugitani H, Oshima Y, Yasumoto T. Isolation and structural elucidation of the causative toxin of the diarrhetic shellfish poisoning. Bull Jpn Soc Sci Fish. 1982;48(4):549–52.

29. Amzil Z, Pouchus YF, Le boterff J, Roussaki C, Verbist JF, Marcaillou-Le Baut C, Masselin P. Short-time cytotoxicity of mussel extracts: a new bioassay for okadaic acid detection. Toxicon. 1992;30(11):1419–25.

30. Proença LO, Schramm M, Tamanaha MS, Alves NE. Diarrhoetic shellfish poisoning (DSP) outbreak in Subtropical Southwest Atlantic. Harmful algae news, IOC / UNESCO. 2007;33:19–20. Available at: http://unesdc.unesco.org/images/0015/001528/152834e.pdf.

31. Haddad V Jr, Moura R. Acute neuromuscular manifestations in a patient associated with ingesting octopus (*Octopus* sp.). Rev Inst Med Trop São Paulo. 2007;49(1):59–61.

32. Cariello L, Zanetti L. α- and β-cephalotoxin: two paralysing proteins from posterior salivary glands of *Octopus vulgaris*. Comp Biochem Physiol. 1977;57(2):169–73.

33. Müller GJL, Lamprecht JH, Barnes JM, De Villiers RVP, Honeth BR, Hoffman BA. Scombroid poisoning. SAMJ. 1992;81:427–30.

34. Morrow JD, Margolies GR, Rowland J, Roberts LJ 2nd. Evidence that histamine is the causative toxin of scombroid-fish poisoning. N Engl J Med. 1991;324(11):716–20.

35. Lerke PA, Werner SB, Taylor SL, Guthertz LS. Scombroid poisoning. Report of an outbreak. West J Med. 1978;129(5):381–6.

36. Boisier P, Ranaivoson G, Rasolofonirina N, Andriamahefazafy B, Roux J, Chanteau S, Satake M, Yasumoto T. Fatal mass poisoning in Madagascar following ingestion of a shark (Carcharhinus leucas): clinical and epidemiological aspects and isolation of toxins. Toxicon. 1995;33(10):1359–64.

37. Louzao MC, Ares IR, et al. Marine toxins and the cytoskeleton: a new view of palytoxin toxicity. FEBS J. 2008;275(24):6067–74.

38. Vasconcelos V, Ramos V. Palytoxin and analogs: biological and ecological effects. Mar Drugs. 2010;8(7):2021–37.

39. Hacon SS, Oliveira-da-Costa M, Gama CS, Ferreira R, Basta PC, Yokota D. Mercury exposure through fish consumption in traditional communities in the Brazilian Northern Amazon. Int J Environ Res Public Health. 2020;17:5269.

40. Santos MC, et al. Outbreak of Haff disease in the Brazilian Amazon. Am J Public Health. 2009;26(5):469–70.

41. Almeida LKR, Gushken F, Abregu-Diaz DR, Muniz R Jr, Degani-Costa LH. Rhabdomyolysis following fish consumption: a contained outbreak of Haff Disease in São Paulo. Braz J Infect Dis. 2019;23(4):278–80.

42. Buchholz U, Mouzin RD, Moolenaar R, Sass C, Mascola L. Haff disease: from the Baltic Sea to the US shore. Emerging Infect Dis. 2000;6(2):192–5.

Infections and Infestations in Aquatic Environments

5

Bacterial and Fungal Infections

Bacterial infections in the skin are suppurate processes that can be primary or secondary (when initial infections occur in other organs but not on the skin) [1–5]. Bacteria or fungi penetrate the skin and the primary skin colonization creates inflammation and suppuration that may lead to hematogenous dissemination with bacteremia and septicemia.

Two Gram-positive cocci are responsible for most of skin and soft tissue infections in aquatic environments: *Staphylococcus aureus* and *Streptococcus* beta-hemolytic group A (Fig. 5.1) [1]. These genera are the most common agents found both in terrestrial and in aquatic environments, and they are responsible for the majority of the infections in humans. Clinical manifestations are local inflammatory process with pain, malaise, high fever, and lymphadenopathy. Clinical exam shows erythematous plaques, with sharply demarcated borders. In later stages, there may be vesicles, blisters (sometimes, hemorrhagic blisters), and even necrosis on the surface of the plaque [1–5].

The infections caused by the *Vibrio* genus can be very severe. *Vibrio vulnificus* is an opportunist Gram-negative coccobacillus, and the infections occur through cuts or trauma in the water (about 30%) or after ingestion of raw or undercooked seafood (1–3 days). People with increased risk of infections by *V. vulnificus* are the immunocompromised (especially patients with liver disease) [5].

Vibrio vulnificus can provoke severe infections with rapid onset. The patient may present with sepsis within hours (in infections by ingestion) or blisters and necrotic ulcers at the point of contact and fever, malaise, secondary sepsis, and septic shock (40–60% mortality!) (Fig. 5.2). The infections caused by *Vibrio* and *Aeromonas* have been defined recently that identify them as the causative agents of necrotizing fasciitis. This severe disease with deep compromise of tissues and systemic manifestations can be associated with S*treptococcus pyogenes* (group A) and, in case of aquatic environments, with anaerobic Gram-negative bacilli and the species *Vibrio, Aeromonas*, and *Clostridium* [6, 7].

Aeromonas hydrophila is a Gram-negative rod/bacillus, which can cause gastroenteritis and severe cellulitis in injuries originated in aquatic environments. *Aeromonas hydrophila* is more common in freshwater environments, but it can occur in estuarine and marine waters.

The infection curses with fast installation. Edema and erythema are the initial manifestations, with high fever, toxemia, bacteremia, crepitation in the tissues, and local hemorrhage (Fig. 5.3). In later stages, it is possible to see blisters, severe cellulitis, fasciitis, and myonecrosis (similar to gas gangrene). Patients with increased risk again are the immunocompromised. Skin infection by *A. hydrophila* is a very serious disease, with high mortality rates (60–70%) [5].

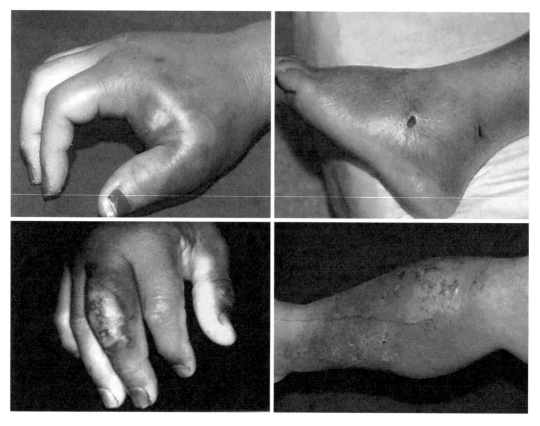

Fig. 5.1 The image shows four infections caused by *Staphylococcus aureus* acquired in aquatic environments. (Photos: Vidal Haddad Junior)

Other infection associated with aquatic environments is erysipeloid caused by the species *Erysipelotrix rhusiopathie.* This bacterium is a Gram-positive organism that provokes cellulites without systemic signs. Epidemiology indicates traumas in butchers and fishmongers and the disease can present with only local manifestations (Rosenbach's erysipeloid) and systemic compromise, which can cause endocarditis. The prognosis is good [1–5].

Pool granuloma is provoked by the *Mycobacterium marinum* species. From an initial trauma in freshwater environments arises an initial nodule that evolves to vegetating-verrucous tumoral lesions and abscesses. Manifestations appear 2–6 weeks after abra-

sions near the bony prominences (generally in the hands) [1–5].

Gas gangrene is a serious infection caused by *Clostridium perfringens* that causes marked local manifestations, deep compromise in tissues, and crepitation in the tissues, and has a high mortality rate. The bacterium *Clostridium tetani* can cause tetanus in wounds originated in aquatic environments. Cholera is caused by one species of *Vibrio*, the *Vibrio cholerae* [1–5].

The possibility of acquiring infections after trauma or contact with seawater or freshwater is reinforced by clinical observation and marine bacteriology: a study shows that it is possible to find highly pathogenic species of bacteria in several locations in aquatic environments, such as

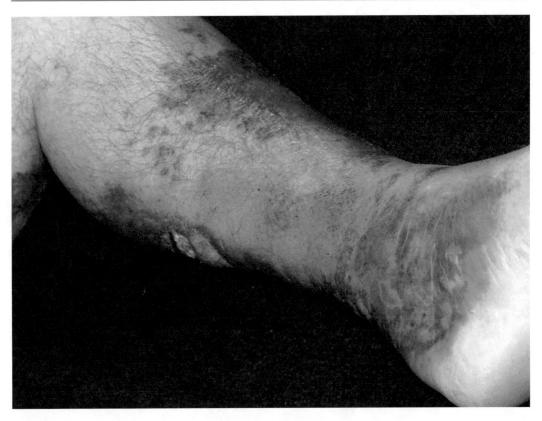

Fig. 5.2 Fast and severe infection with clinical features of necrotizing fasciitis is suggestive of *Vibrio vulnifucus* infection, with high levels of death. (Photo: Vidal Haddad Junior)

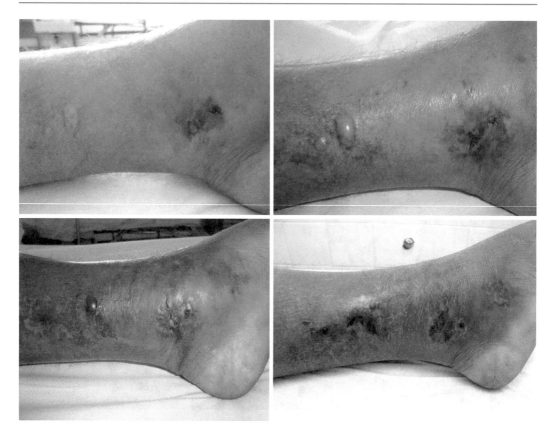

Fig. 5.3 Infections caused by *Aeromonas hydrophila*, as the *Vibrio* infections, are extremely severe. Both are acquired in aquatic environments, but *Aeromonas* is more associated with freshwater environments. This infection came after a sting caused by a freshwater stingray. (Photos: Vidal Haddad Junior)

shells (*Vibrio alginolyticus*), sponges (*Vibrio alginolyticus, V. cholerae,* and *V. vulnificus*), submerged rocks (*Vibrio alginolyticus*), corals (*Pseudomonas putrefaciens* and *Vibrio* sp.), sea urchins (*Vibrio parahaemolyticus* and *V. vulnificus*), hydroids (*Vibrio alginolyticus/vulnificus*), seawater (*Vibrio* and *Pseudomonas* sp.), triggerfish *(Pseudomonas putrefaciens* and *Klebsiella pneumonia)*, and shark's teeth (*Pseudomonas putrefaciens, Aeromonas hydrophila, Vibrio algynolyticus,* and *V. vulnificus*). Recent studies show the presence of *A. hydrophila* in stingers of freshwater stingrays and other fish and animals of freshwater environments [8–10].

The presence of fungi in aquatic organisms has been confirmed when sporotricosis that occurs post-trauma was reported in fishermen who were injured by fish. Recent studies have shown the presence of pathogenic fungi in traumatogenic structures and skin of freshwater fish such as piranhas, *Tilapia* sp., and others (Figs. 5.4 and 5.5) [11, 12].

Fig. 5.4 The Tilapia fish were introduced in various countries, due the good meat and facility of creation. The rays of the fins are very sharp and cause wounds in fishermen. (Photo: Vidal Haddad Junior)

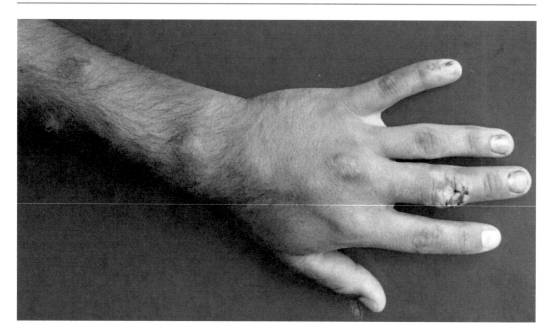

Fig. 5.5 Fungal infections also can occur after trauma with fish. In this case, the patient presented a sporothricosis after a perforation in the third finger of the left hand that shows typical ascendant lymphangitis. (Photo: Vidal Haddad Junior)

Treatment

Staphylococcus/Streptococcus Penicillin is still the treatment of choice for mild infections. The initial approach is use of oral cephalexin (500 mg every 8 hours), oral clindamycin (450 mg 6/6 hours), or oral ciprofloxacin (750 mg every 12 hours). In cases of moderate infections, it is advisable to use cephalosporin (cephalothin EV 1–2 g of 6 in 6 hours) or amoxicillin + clavulanic acid (500 mg EV 6/6 hours). Severe cases should use clindamycin in combination with an antibiotic such as ciprofloxacin [1].

The infections by *Vibrio vulnificus* are treated with doxycycline 100 mg IV or orally twice a day AND ceftazidime 2.0 g IV 8/8 hours or ciprofloxacin 750 mg orally or 400 mg intravenously twice a day. The infections by *Aeromonas hydrophila* are treated with ciprofloxacin 500 mg IV 12/12 hours. Sulfamethoxazole and trimethoprim, and doxycycline or chloramphenicol are optional [5].

Erysipeloid is treated with penicillins, ampicillin, cephalexin, or erythromycin. Pool granuloma (*Mycobacterium marinum*) is treated with a combination of two or three drugs – clarithromycin, ethambutol, and rifampin – for months [1–5].

Injuries from aquatic environments, even small and superficial, should be intensively cleaned with soap and water. Use of topical antibiotic (mupirocin or fusidic acid) is important. If there is obvious infection, care must be twofold. Remember that serious infections manifest themselves after days already in your care! Lacerated/deep/extensive injuries must be intensively cleaned and systemic antibiotics should be used (cephalexin, for example) for 10 days.

Causes of wounds that can present with infections in aquatic environments are as follows: bites, stings, punctures by rays of fins, spines of sea urchins, seawater, coral cuts, and injuries that occur during practicing water sports such as those caused by hooks, surfboards, boat propellers, or jetski.

Worm Infestations After Consumption of Aquatic Animals

The increased consumption of raw or undercooked fish and seafood has been disseminating infestations by worms that were more observed in countries that have always had this habit, such as Japan, Peru, and Chile. The incidence of these diseases is 100 cases per year in Japan and lower in other countries, with a death rate of approximately two cases per year [4, 5].

These worms have always existed in the countries that are now affected by these diseases, and the growing infestations are highly linked to changing eating habits, especially consumption of raw fish. Due to the demand for freshly caught raw fish, the process of freezing is skipped and hence the larvae in the fish are not destroyed.

Consumption of sashimi and ceviches is healthy, but it involves some risks,. For example, larvae of worms in fish can infect humans and, although they do not complete their life cycle in humans, they can cause skin and gastrointestinal problems when they move.

The incidence of these infestations is on the rise and the health teams should be aware of the signs and symptoms, as they are difficult to interpret, and require a well-documented patient history, where information on the consumption of raw fish meat is essential.

Nematodes or Roundworms

Nematodes are round-bodied worms capable of causing serious consequences in the human digestive tract, when ingested in the form of raw or undercooked fish meat. The most important worms that undergo their parts of life cycle in the musculature of fish are anisakiasis, eustrogilidiasis, and gnastotomyiasis.

Anisakiasis

Anisakiasis is caused by the worms *Anisakis simplex*, *Pseudoterranova decidiens*, *Contracaecum* spp., and *Hysterothylacium* (*Thynnascaris*) spp. (Fig. 5.6). The worm completes the adult stage of its life cycle in marine mammals, such as seals, sea lions, and dolphins. These animals eliminate the parasites' eggs in their feces, which turn into larvae in crustaceans and fish. These larvae would complete their life cycle in other marine mammals but may end up in accidental hosts such as humans. The point where the larva attaches to the mucosa can be seen as a small whitish nodule, visible when performing an

endoscopy or laparoscopy to remove the larvae. This is the best form of treatment, which can still be tried with the use of albendazole 400 mg twice a day for 5 days [13].

Eustrongylidiasis

Eustrongylidiasis is caused by the nematodes *Eustrongylides tubifex*, *E. ignotus,* and *E. excisus*. Adult worms are parasites of birds that feed in aquatic environments. The worms' eggs are carried via the birds' feces and the larvae can settle in oligochaetes, reptiles, and fish until they are ingested and complete their life cycle in the birds.

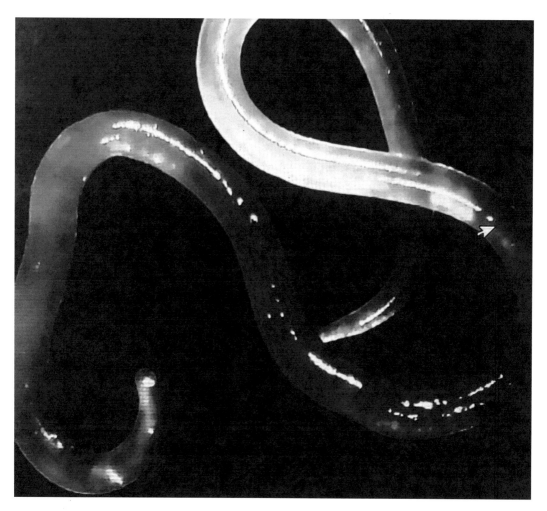

Fig. 5.6 *Anisakis* sp., a roundworm found in fish meat and capable of causing accidental infestations in humans. (Drawing: Ariadne M Haddad)

Occasionally, they settle in humans, where they do not complete their life cycle [13]. Ivermectin, at a single dose of 200 mcg/kg orally daily, for 2 days, is the medication of choice for this worms.

Gnathostomiasis

Gnathostomiasis is caused by the helminth nematode *Gnathostoma* spp. (Fig. 5.7) acquired by the consumption of raw or undercooked aquatic animals contaminated with larvae of *Gnathostoma,* which is the third stage of its development (AL3). The larvae will never become adult worms in human host. The disease is endemic to Southeast Asia and Central and South America. Humans are occasional hosts and the larvae can cross the digestive tract, subcutaneous tissue, and other solid organs. Skin gnathostomiasis occurs 3–4 weeks after larvae ingestion, and a migratory skin lesion ranging from a nodule to an infiltrated ill-defined

mass appears with peripheral and local eosinophilia (Fig. 5.8). Treatment with albendazole or ivermectin is effective, but some patients may experience recurrence even after therapy [14].

When in the human body, the nematodes larvae settle in the digestive tract, but they cannot complete their life cycle and are expelled through the feces after about a month. In the meantime, especially in the first few days, they can cause non-specific symptoms, such as nausea, vomiting, and epigastric pain. They can also simulate acute abdomen and rarely cause perforation of the intestinal wall, by fixation on the mucosa muscles. Eosinophilia arises at the site and in the bloodstream and can be associated with allergic processes such as hives (suspect when a person is allergic to different fish!). The involvement of the spleen, liver, and lungs is even more rare.

The aquatic organisms most suspected of transmitting nematode parasitic infestations are the herring (*Clupea harengus*), the cod (*Gadus*

Fig. 5.7 *Gnathostoma* sp. is a worm that infests humans accidentally and causes important lesions on the skin and other organs, due to the migratory nature of the larvae. (Drawing: Ariadne M Haddad)

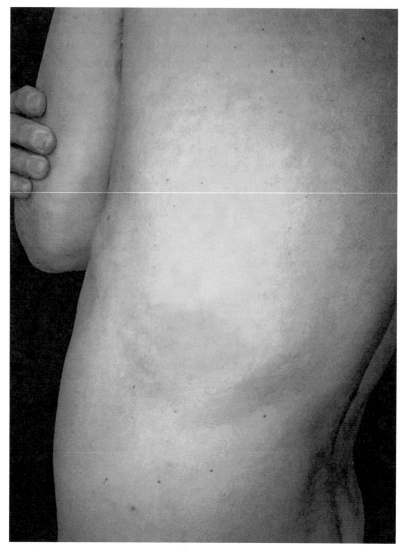

Fig. 5.8 This patient has an infiltrated and erythematous mass on the left flank, in a verified case of gnathostomiasis after consuming raw meat of peacock bass in an Amazonian River. (Photo: Vidal Haddad Junior)

morhua), the tuna (*Thunnus* sp.), and the squids (various genera and species). In freshwater environments, consumption of the tucunarés or peacock bass (*Cichla* spp.) and traíras (*Hoplias malabaricus*) is the main causes of the infestations (Fig. 5.9).

As a general rule, boiled or roasted fish do not transmit the diseases, but if the option is to eat raw or undercooked meat, the fish must be frozen at −35°C or less for at least 1 day. If frozen at −20°C, 7 days of freezing is required.

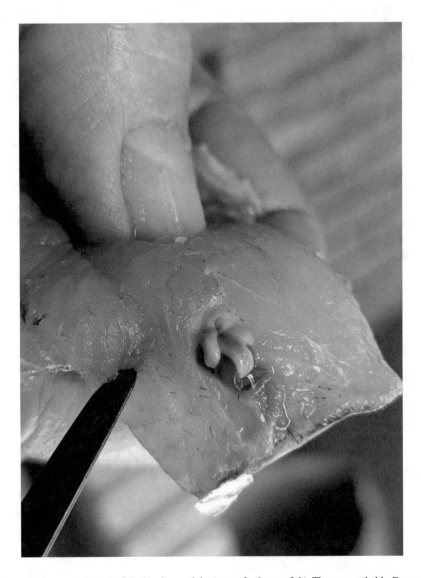

Fig. 5.9 Worms in the meat of a traíra fish (*Hoplias malabaricus*, a freshwater fish). They are probably *Eustrongylides* sp. (Photo: Vidal Haddad Junior)

Flatworms

Diphyllobhotriasis

Diphyllobhotriasis is a fish parasite also transmitted by the consumption of raw or undercooked fish. The worms responsible for the infestation are of the genus *Diphyllobhothrium*, with *D. latum* being the most common species. Salmon (various genera and species) is the fish most associated with the infestation. The worm can reach 10 meters and parasitize humans for years. It occurs more frequently in East and Southeast Asia due to tourism and consumption of oriental food.

Diphyllobhotriasis cases have been reported due to the consumption of sushi and sashimi. In 2005, 45 cases were diagnosed in the laboratory to be infected by *D. latum* in the State of São Paulo, Brazil, with 25 cases eliminating fragments or the entire parasite. All of these cases were associated with the consumption of raw salmon. The cycle requires two intermediate hosts (a copepod crustacean and a freshwater or saltwater fish). The larva found in fish that infests humans is called plerocercoid, which in the human intestine turns into an adult worm (usually just one per human) and releases proglottid eggs into the host's feces. The symptoms may not exist or be non-specific, including malaise, nausea, and vomiting, but there may be diarrhea, abdominal pain, and megaloblastic anemia due to malabsorption of vitamin B12 by the host [13].

The preferential treatment of diphyllobhotriasis is carried out by means of a single oral dose of praziquantel – 5–10 mg/kg.

References

1. Lupi O, Belo J, Cunha P. Routines for diagnosis and treatment of the Brazilian Society of Dermatology. Rio de Janeiro: Editora Guanabara-Koogan; 2012.

2. Habif TP. Clinical dermatology: a color guide to diagnosis and therapy. 5th ed. New York: Mosby; 2009.

3. Sampaio SAP, Rivitti E. Basic dermatology. São Paulo: Artes Médicas; 2007.

4. Bolognia JL, Jorizzo JL, Rapini RP. Dermatology. USA: Elsevier; 2003.

5. Scott S, Thomas C. All stings considered: first aid and medical treatment of Hawaii's marine injuries. Hawai'i: University of Hawai'i Press; 1998.

6. Edlich RF, Cross CL, Dahlstrom JJ, Long WB. Modern concepts of the diagnosis and treatment of necrotizing fasciitis. J Emerg Med. 2010;39(2):261–5.

7. Vayvada H, Demirdover C, Menderes A, Karaca C. Necrotizing fasciitis: diagnosis, treatment and review of the literature. Ulus Travma Acil Cerrahi Derg. 2012;18(6):507–13.

8. Haddad V Jr. Cutaneous infections and injuries caused by traumatic and venomous animals which occurred in domestic and commercial aquariums in Brazil: a study of 18 cases and an overview of the theme. An Bras Dermatol, Rio de Janeiro. 2004;79(2):157–67.

9. Millington JT, Wilhelm P. Marine microbiology of Roca Alijos. J Wilderness Med. 1993;4:384–90.

10. Haddad V Jr, Cardoso JLC, Garrone ND. Injuries by marine and freshwater stingrays: history, clinical aspects of the envenomations and current status of a neglected problem in Brazil. J Venom Anim Toxins Incl Trop Dis. 2014;19(1):16.

11. Haddad JRV, et al. Cutaneous sporothricosis associated with a puncture in the dorsal fin of a fish (*Tilapia sp*): report of a case. Med Micology. 2002;40:425–7.

12. Leme FCO, Negreiros MMB, Koga FA, Bosco SMG, Bagagli E, Haddad V Jr. Evaluation of pathogenic fungi occurrence in traumatogenic structures of freshwater fish. Rev Soc Bras Med Trop. 2011;44:182–5.

13. Taira KK. Principais parasitas com potencial zoonótico transmitidos pelo consumo de pescado no Brasil. (Main parasites with zoonotic potential transmitted by fish consumption in Brazil). Monograph. Specialization Course in Management in Agricultural Defense: Emphasis on Inspection of Products of Animal Origin, Sector of Agricultural Sciences, Federal University of Paraná.Curitiba, 2011. 39 pp.

14. Haddad V Jr, Oliveira IF, Bicudo NP, Marques MEA. Gnathostomiasis acquired after consumption of raw freshwater fish in the Amazon region: a report of two cases in Brazil. Rev Soc Bras Med Trop. 2020;53:e20200127.

Index

Printed in the United States
by Baker & Taylor Publisher Services